Aha… That is Interesting!
JOHN H. HOLLAND,
85 Years Young

Exploring Complexity

Series Editor: Jan Wouter Vasbinder
Director, Complexity Program
Nanyang Technological University

Editorial Board Members:
Geoffrey B. West, *Santa Fe Institute*
John Steve Lansing, *Nanyang Technological University*
Robert Axtell, *George Mason University*

For four centuries our sciences have progressed by looking at its objects of study in a reductionist manner. In contrast complexity science, that has been evolving during the last 30–40 years, seeks to look at its objects of study from the bottom up, seeing them as systems of interacting elements that form, change, and evolve over time. Complexity therefore is not so much a subject of research as a way of looking at systems. It is inherently interdisciplinary, meaning that it gets its problems from the real non-disciplinary world and its energy and ideas from all fields of science, at the same time affecting each of these fields.

The purpose of this series on complexity science is to provide insights in the development of the science and its applications, the contexts within which it evolved and evolves, the main players in the field and the influence it has on other sciences.

Aha... That is Interesting!
JOHN H. HOLLAND,
85 Years Young

Editor

Jan W. Vasbinder

Nanyang Technological University, Singapore

World Scientific

NEW JERSEY · LONDON · SINGAPORE · BEIJING · SHANGHAI · HONG KONG · TAIPEI · CHENNAI

Published by

World Scientific Publishing Co. Pte. Ltd.
5 Toh Tuck Link, Singapore 596224
USA office: 27 Warren Street, Suite 401-402, Hackensack, NJ 07601
UK office: 57 Shelton Street, Covent Garden, London WC2H 9HE

British Library Cataloguing-in-Publication Data
A catalogue record for this book is available from the British Library.

Exploring Complexity — Volume 1
AHA.... THAT IS INTERESTING!
John Holland, 85 Years Young

ISBN 978-981-4619-86-8

Tribute to John Holland

I got to know about John when Jan initiated Institute Para Limes (IPL). In fact John, Jan and I are co-founders of IPL. Then I got to know John through his books, "Hidden Order" and "Emergence", and his work at the Santa Fe Institute. I started to realize that John is one of those very few scientists, who all by themselves and by their pursuits change the course of science, and the treasure of human knowledge. There is hardly a field in science or problems for humanity that is not affected by John's work on complexity and, more in particular, complex adaptive systems. Whether it is the immune system, the brain, ecology, economy, language acquisition, sociology, or the understanding of what make inner cities tick, John's work and insights have changed the way in which science seeks to understand the workings of our world and the systems it is made of, and are setting directions for current and future research.

So it was with great pleasure that Nanyang Technological University offered to host the conference, organized by Institute Para Limes and the Santa Fe Institute in honor of John's 80th birthday, in 2009. In that conference, *Adaptation, Order and Emergence*, the tremendous impact of the science that John Holland spawned on a great variety of scientific questions and societal problems was shown in an impressive way by some of the most distinguished scientists of our time. John himself gave an enlightening talk about his work, the direction in which he is now moving, and the relevance thereof.

That conference was the first time I actually met John. I have met him many times since. I have come to understand why this modest, very humorous and friendly man has become such a giant in science. He is an example to us all and a source of inspiration.

So I am honored to have been asked to write this little note about John for the book that his friends Jan, Helena and Jing (all working at NTU) have put together for his 85th birthday.

Happy birthday John and thanks a lot.

Bertil Andersson
President Nanyang Technological University

1

Preface

As Bertil Andersson's tribute testifies, John Holland is a special person. So special that about 30 of his friends wrote a book for him to celebrate his 85th birthday. Helena Hong Gao, Jing Han and I organized it. The book was presented to John in Ann Arbor on his birthday. It was a personal gift by his friends to be distributed by John at his discretion.

John liked the book. So much so that he agreed that it be edited for a wider readership. Helena and Jing delegated that task to me.

The book you hold in your hand is the result. It differs from the original edition only to the extent that more of John's friends contributed and that some minor editorial changes were made in the original contributions.

The book is richer for it, showing even more aspects of John.

This is not a book of science. It is a book about a scientist who changed the way science is done, about an eternally young and explorative mind, a great teacher and inspirer, a family man and a friend. It is a book about a modest, humorous and warm person. And it is very personal in the sense that this book could only be written by family and friends, who know John well.

John is 85 years young. He is not immortal and neither are his friends. One day books will be written about how John changed science and how that science changed and continues to change how we understand our world. Maybe the writers of these books will have known John, maybe not.

I trust that in researching for their books, they will find the connection between the richness of John's science and the warmth of his person.

Maybe this little book will contribute to that.

Jan W. Vasbinder
Singapore, August 2014

Contents

1   Message to John

John,

This is a very special book.
It is a book for you written by your friends.
It is a book for your friends, inspired by you.

You may have an idea what your friends mean to you, and after reading the book you may change your opinion about some of them. But we think you may have only a limited idea what you have meant and mean to them. This book may change that too.

Your friends who wrote this book are of course only a selection of your friends. They are the ones that we managed to find and who reacted to our call for contributions. In the letter we sent we asked each one for:

A short story (5 pages or less) about him (you). It should be a personal story that only you (the friend) can write. The combined stories should:

- *Highlight aspects of the becoming of his (your) science that will most likely not be highlighted in the many books about John's (your) science that undoubtedly will be written in the future.*

- *Bring out anecdotes, stories and insights that otherwise will get lost. Stories about his quests, his collaborators, the worlds he lived in, the changes he observed and was part of, his expectations, etc..*

- *Make visible his incredible mind, his boyish curiosity and explorative energy, his philosophy of life, how that developed and how he brings that into practice, including his enormous hospitality and natural inclination to make friends.*

As you see, we asked for totally non-biased contributions. So you may consider this book as a fair presentation of you. We do not know to what extent you will be comfortable with this presentation. But there is no escape: this is how your friends see you when they think about you. And all of them were only too happy to share that.

In great friendship,

Jan, Helena, Jing,
Singapore, 2 February 2014

P.S. We have included all the contributions we have received. We did not edit the texts. We only corrected a few typos and formatted the whole book.

2   Shirley Holland Ringgenberg

~ for my brother, John Henry Holland

The rest of the world knows me as Holli Ringgenberg, my brother calls me Cis which is short for Cistern......taking into account John's sense of humor, I'll let your imagination suggest how it came about that he started calling me Cistern.

John and Holli

The year John was (almost) two, I was born and our folks built a summer home on Clear Lake in northeastern Indiana. We were fortunate enough to spend every summer there, no matter where we were living in the winter; by the first of June, we arrived at Clear Lake. Friends that we made

My darling brother saved my life. While the folks were working on the garden along the side of the house; he yelled up to them, "You won't let me go swimming, but Shirley is."

Of course I had fallen off the seawall — Daddy rolled me over the proverbial barrel and brought me back. If John hadn't called up to them I wouldn't have survived!

at the lake, and grew up with, we stay in touch with and still see occasionally today, in the Winter of our lives.

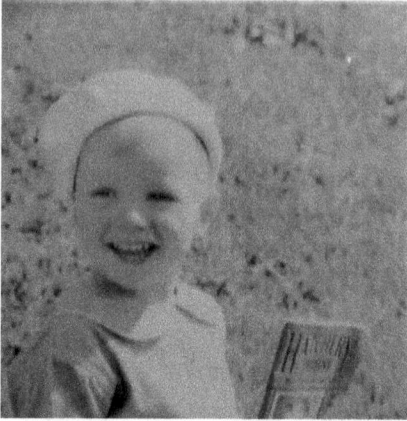

The first thing I recall hearing about John's 'boyish curiosity' was probably when he was three. He had gone with Daddy to fill the gas tank of our car, a very large Packard. It looked like such a neat thing to do that when they got home he decided to fill it again. So he did. With sand. There being no water around, much less a gas pump, he'd resorted to our sand box. Now this Packard wasn't just any old Packard but top of the line; the fuel system was welded to the interior and the dealer had to resort to hoisting the car in the air, upside down, in an attempt to flush the sand out. The car never ran quite right again.

Our folks raised us in their "old school manner, ie. too much affection lavished on a child results in an irresponsible adult". I believe this mirrors the manner in which they were raised and, consciously or unconsciously, they followed that same pattern. Nevertheless, while we may have been deprived of some things, we never wanted for anything; and I can remember only one spanking ~ and John was on the receiving end of that one.

Our parents obviously did things right — we have both led successful lives, albeit in completely opposite directions: John, into academia; I, into business. You're aware of my brother's successes; in tribute to our parents, may I tell you that in the mid-1980s, while I was still in corporate life working for a Fortune 50 company as a Regional Manager, I was the highest ranked female in the whole corporation. (Today, that organization boasts a female CEO.)

John was always the studious one (his bedroom had a DESK! Mine simply had a dressing table.) I can remember my first appearance at Van Wert High School and meeting Benny Spieth, the science-chemistry-physics teacher. He said, "Oh, you're John Holland's sister, we'll be expecting big things from you." Uhhh yeah, right — just not exactly the things you were probably expecting, Mr. Spieth! In my junior year, taking same said Mr. Spieth's physics class, I was totally lost. I couldn't wait for my big brother to come home from MIT

In high school chemistry class John and a friend wanted to see how the ventilation system worked. Whatever they let loose, closed the school down for the rest of the day. Please understand that since I'm two years behind John a lot of what I remember is gossip (?). I do think it's a fact that although he was slated to be Valedictorian of his (high school) class, one of his stunts caused him to be suspended for the last term of his senior year. Our folks never knew, he left every morning just on time. I never knew either........and he still got a (full?) scholarship to MIT at Christmas — he was majoring in nuclear physics and would explain it all to me. He came home, he explained it all to me. I didn't understand a word he said then and he's probably hard pressed to get me to understand his chosen profession to this day.

One of the most important things instilled in us was that there isn't anything we can't do, nor anything we can't accomplish — we simply had to want it enough to put in the necessary effort and to persevere. Both our parents expected us to do well, had faith that is what would happen, and believed in us. (And I did pass Mr. Spieth's physics class.)

A lot of what we learned was modeled after the example they set. Proper manners for instance. And from my niece Gretchen, (John's eldest) I discover that John's youngsters learned their good manners from following the example set by the parents. I do remember, when we were probably three and five years old, the couplet Mother used to teach us how to eat soup properly, "I push my spoon out to sea, then slowly bring it back to me."

By no means, were we model children. Our folks started taking us to restaurants early on — a learning experience certainly. One time, probably in the same timeframe as the soup lesson, we were seated in a restaurant, Mother and Daddy had probably gone to hang up coats. John and I slipped out of our seats and went from table to table, peering over the edge of each, to see what others had on their plates. Now that piece of the 'learning experience' had to have come from my brother's creative mind.

As we grew up, we ran in totally different circles, both at the lake and at home (where ever home might happen to be that winter). Daddy was the classic and successful entrepreneur. He started his career as advertising manager for Wayne Feeds out of Fort Wayne, Indiana, officed in the Board of Trade building in Chicago. We lived in Winnetka, a Chicago suburb. John had been born in Fort Wayne, so Mother, being the strong-minded woman she was, decided she wanted the same physician to deliver me. Mother, Daddy and John boarded the train to Fort Wayne and stayed in a rental apartment until I was born. Then we all returned to Winnetka on the train, with a nurse in tow. In those days, new mothers usually stayed in the hospital at least ten days after giving birth.

Daddy took what he had learned about the manufacture of feed for farm animals and opened Old Fort Mills in Marion, Ohio, in 1934 (or 1935 maybe, I'm too old to remember such detail). Two or three years later, needing larger facilities, he found what he wanted in Piqua, Ohio, opening Holland Mills. Three years later, Holland Mills burned to the ground; the fire was set by a disgruntled ex-employee, discharged for drinking on the job. He set the fire to "get even". However, when he found himself behind bars for arson, he tried to get even again, claiming that Daddy had paid him to set the fire. John was in the sixth grade, I was in the fourth, a very grim time for our family. Daddy was indicted and, at the end of the trial, it took the unanimous jury twenty minutes to declare him not guilty.

John's father making a face

Once again, Daddy had to start over. He chose Ohio City, Ohio, a tiny crossroads with a lot of railroads. We moved to Van Wert (seven miles from Ohio City), a town of just-under ten thousand, where John and I both started, and subsequently graduated from VWHS. Daddy built the manufacturing plant, the elevators, the office building and thus, Holland Pioneer Mills was born. In addition, he built a soybean processing plant and was a pioneer in the extended uses for soybeans. For this he was widely recognized and honored. For many years, Daddy had been involved with Rotary International; he continued as an active Rotarian in Van Wert. When he finally sold the mills and retired, Rotary bestowed on him a seldom given honor, making him a lifetime Rotarian, giving him full access to any Rotary Club, any where in the world. In those days, (late 1940s) Rotarian membership was by invitation only, limited to men, and limited to two men of any one particular profession.

Our Father also was an active supporter of Boy Scouts, both as a leader and a financial supporter. Nothing made him prouder than the day John achieved Scouting's highest rank, Eagle Scout. And then later, the Order of the Arrow. As a reward — or perhaps it was the incentive — at sixteen years old, my brother started flying lessons.

Which leads us to our Mother. She was certainly a woman before her time.....she got her pilot's license at Wright Pat when we still lived in Piqua. She worked outside the home at a time when women simply didn't hold a job unless it was an absolute necessity. In Ohio City she was the corporate accountant for Holland Pioneer Mills. She loved to travel. Daddy, not so much. After John and I were older she'd take off for Hawaii, or Australia, or Alaska, or around the world — and always by ship. When she was in her 50s (I'm guessing again), Wolf and Dessauer, Fort Wayne's version of Saks Fifth Avenue, asked her to model for them. Which she did, and did with a lot of class. Mother was an avid bridge player, mostly duplicate, and very good. John's love of gaming certainly came from her.

In Van Wert I can remember him creating games with his friends, out of paper and huge sheets of cardboard; then, more substantial board games using lead soldiers he himself made in our basement with his soldering set. His circle of friends during his high school years was small; but, again, their unwavering loyalty to each other has lasted a lifetime.

John's continuing interest in mushrooming was triggered by — whom else? — our Dad, at — where else? — Clear Lake. His love of sailing started early too; we were probably eight and six when our parents bought a 16 foot racing dinghy that year. Again, we went in different directions ~ he went on to teach his kids how to sail and passed on his love of sailing. He still has a sailboat at the house in the U P. I haven't been sailing in 60 years.

Gretchen reiterated the story of their trek to the Green Sand Beach on the Big Island of Hawaii (where I had taken up residence to raise my three youngsters.) Gretchen said it seemed like they walked miles and miles — and they had. As I recall it's a little over three miles from the access road to the green sands. Perseverance instilled in John early on, and again, passed on to his children.

Gretchen's description on her Dad's goofy side and his fun-loving quips is a perfect description on one facet of my brother — 'Cistern' is the product of one of his goofy quips.

A quirky sense of humor: When John was an undergrad at MIT, somebody from the 'powers that be' declared that coat and tie would be required at the evening meal. My brother and all his cronies showed up in the required 'coat and tie'...and nothing else! Probably trousers but little else.

The time line in the edition of this tome states that John received the first Ph.D. in Computer Science. The story this doesn't tell is that it took months (and months?) to put together the team to give him his oral examination. I've often wondered if they were looking for professors who knew more about Computer Science than my brother, or did they just settle for people who probably knew as much???

One thing a lot of folks don't know about my brother is that he teaches by example, and uses words as a last resort. I love John dearly, am very proud to be his sister, and continue to stand in awe of his knowledge and his accomplishments.

Mother and Daddy would be proud.

Shr

3   Marianne Ringgenberg

Dear Uncle John,

When Mom told me about this book being compiled I wrote to Jan and asked if I could add a few words. The thing I want to say is simply this, you have been a constant in my life, in a topsy-turvy kind of life. You in that solid house in A2 have been there if I needed a place to land. We have shared wonderful times — dinners at the Hau Tree Lanai Kaimana Beach Hotel, Honolulu, one memorable night in particular when toddler Manja was in rare form, your patience and humor were on display in all their glory. Meeting several times at the World Economic Forum in Davos Switzerland — I was in awe of the topics you spoke of and the audience you spoke to. A couple of New Years in Wisconsin. Visiting you up at Gulliver, the snow mobile rides to the restaurant instead of cars. Small kid times, at the Green House in Pahoa, the stick fish I was so afraid of but you made me see they were just fellow creatures. The picture I include must be one of the few with You and Mom and all us cousins. It was 1982, the Christmas after Dad died and a more solid rock you could not have been.

I am always proud to display your books, and let people know right off the bat the way you can converse with anyone about anything — case in point a very down to earth conversation we had once on just how complex ants are. Who'd a thunk it. Well you did of course.

I am proud to count myself among your family and love you dearly.

Marianne, also known as the eldest niece. ;-))

4   Gretchen Holland Sleamon

My Memoir of John Holland, known to me as Daddy

What I have to thank my Dad for:
My wacky sense of humor and love of bad puns
Love for the outdoors and knowledge of flora and fauna
Love of nature
Love of adventure
Love of anything JRR Tolkien and reading
My sweet tooth
Love of travel
Confidence
Ability to talk to anyone/outgoing nature
Love of Music
Manners
Mathematics ability
Curiosity for how things work
Appetite for trying new things

My Dad has always been a teacher, even before he became a professor. I see delight in his eyes as he introduces something he finds fascinating to someone, when showing others anything new. As a very young child I remember loving to go up to his "office", the attic in the old farm house we rented, I don't even think it had heat up there....but it was where he worked. I was mesmerized by his work space. I recall being shown his Computer lab at University of Michigan. The computer was the entire room. He showed us how you could play a simple tennis-like computer game called Pong.

When I was almost five, my parents built the house he lives in today. I went with them and "helped" plant hundreds of little pine saplings that are now a dense full grown pine forest. I thought it was great at first and then got pretty tired and whiney. At age 5, the interest died out rather quickly as I realized this was actual work.

My Dad loves flying kites from the hill top that the house sits on overlooking the Huron river valley. He would get it way up in the sky and let us fly it, too.

We regularly went on nature walks, whether it was walking the paths he mowed on our 15 acres, or hiking way over to the natural marsh on the "cows grazing land" next to our property (with his reassurance that the ferocious, humongous cows would not bother me). He always taught my sisters and I about different plants and animals during these excursions. We would go on adventures out in torrential rain to watch the water raging down the flooded gully which was normally a small stream that meandered to a pond in our property. The pond was always there, but my Dad put in some fill dirt at one end of it where the water ran out to make it larger. At one point he also experimented with what he named "the waterworks" which were a series of trenches he dug to divert more run-off to the pond. The pond got very big that first year of the waterworks! We were always watching how things changed through the seasons and different paths the streams would follow to the pond. Once a tornado went through the grazing fields near our house and we could see large branches flying through the air. My sister and I were terrified. We had our blankets and favorite stuffed animals and were telling Daddy he had to go to the basement with us. He was up in his study fascinated by the unique sight, his curiosity about how things work, piqued. It was likely he knew it was not going to blow our house away and I think he did go to the basement with us eventually.

Discovering different kinds of mushrooms and identifying them based on spore patterns and photos was always a favorite pastime. Frying them up and trying them out when they were deemed to be safe and non-poisonous was added adventure. My Dad loves to try new foods, and I have learned that love as well. I think my Dad's favorite dessert is still the sour cream hickory nut frosted yellow cake that my Maternal Grandma used to make. He has requested it for special occasions, and I have made it. I did substitute walnuts for the hickory nuts that my Grandpa used to harvest and crack, getting the meats out by hand for my Grandma.

My Dad read to us regularly. I love the way he reads…with different voices and great expression. J.R.R. Tolkien was a staple in our house. Such vivid pictures were painted in our minds, that the characters seemed real. My Dad, sisters and I were amazed at how closely the characters resembled the images in our minds when the movies came out. We all went to see them together. My Dad taught me to play cards, board games and tennis. He loves games and almost always wins! I don't recall ever beating him at tennis. As we were not a particularly religious family, Sunday mornings for us meant salt sticks from the Blue Front and Tchaikovsky playing on the stereo in the study. My Dad listened to all the hip music of the time as well. Judy Collins, Joan Baez, Cat Stevens, Nancy Sinatra, Peter, Paul & Mary, the Clancy Brothers to name a few, were favorite vinyl records we regularly played. He took us to plays, the ballet, concerts, the movies, and to fancy restaurants. We were taught good manners and were expected to use them. Most of the time we were well behaved, but it only took a glare, "the look", from Daddy to know when you were stepping on the line. He was strict, but in good way. We didn't talk back (when we were children) or disrespect, but we did voice our opinions. We were exposed to so many different cultural experiences and we thought that this was how everyone grew up.

For many years we have vacationed on the South shore of Lake Michigan in the Upper Peninsula each summer. We'd walk the miles of pristine beach looking for interesting rocks, driftwood, animal tracks (which we would go back and identify) and sea glass. My Dad loved to show us things like a deep, fresh, spring fed lake, rare species of birds, seldom seen animals, and these are the things I now love to show my kids. He still delights in showing these things to visitors of our summer place in the UP. We'd wait for the days when the waves were huge on the Lake MI shore so we could learn and practice body surfing — even if it was only 60 degrees out. Daddy was always a good sport and got in the water with us. Learning to sail with my Dad was exciting and a bit scary, the boat tipping so far over in the wind that I was sure it would capsize every time... and sometimes it did. My sister and I were sure some of the accidental tipping was on purpose, but all a part of learning what to do.

Hawaii was a huge part of my upbringing. John's sister, my Aunt, and my cousins lived there as well as his parents, my grandparents. We had so many wondrous times fueled by my Dad's curiosity and the wealth of local knowledge gleaned from my Aunt Shirley. I went to school there while my Dad was on sabbatical. The first few trips there, we stayed in a huge old lodge that had an outhouse as the bathroom with a big open wooden water tank next to the kitchen filled with rainwater. There was a long open air breezeway from the living room and kitchen area to the bedrooms area, of which there were many to choose from. There were lime and lemon trees right outside the kitchen window and all manner of unusual fresh fruit just growing everywhere. This gem of a palace, painted all green, with acres of tropical splendor, lined with coconut palms along the jagged ocean shore will forever to be remembered by us as the "green house". It had a large tidal pool that had a small bridge at the mouth where the sea rushed in at high tide, and drained almost completely out at low tide. It was so exotic, yet completely normal for us to walk way out on the lava reef when the tide was out and discover all sort of sea creatures and crabs, an of course identify all of them. My Dad taught me to swim in the salt water of the Pacific Ocean, with the added confidence that "salt water makes you more buoyant". We ventured through lava tubes at low tides, rode the waves coming into the tidal pool at high tides....sometimes there would even be small whirlpools as the water rushed in from a large wave. We learned not to be afraid of the fish. My Dad assured us, by comparison we were so much bigger, and they were "just being nosy" when they swam near us. Looking at them through a mask while snorkeling, we thought they were interesting but still did not like them coming too close.

My Dad was always very daring in our eyes, taking the rugged roads that said "Closed" in search of new and interesting sights, going past the sign at the lava cliffs that said "keep back" to get a better view of the surf crashing in, venturing all the way down a closed road to where the still steaming lava flow went right across it. I remember walking what seemed like miles down a dusty dirt road to find the green sand beach. We kept asking "how much further?" but then the amazing sight made us completely forget the trek to get there. It was a natural cove with cliffs of olivine that had become green crystalline sand. We would hike over the rough lava fields in search of a seedling starting new life on the barren land. We went everywhere passable on the Hawaii Volcanoes National park and saw molten lava bubbling and spouting in the caldera. We hiked over boardwalk paths by steam vents and old cinder cones from previous eruptions. It was exciting and seemed a bit dangerous. I now realize we were never in any jeopardy, just very lucky to have such an adventurous leader.

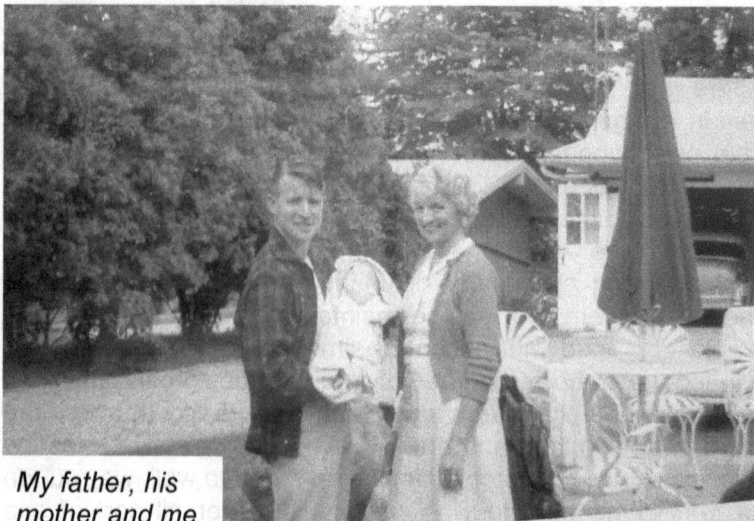

My father, his mother and me

Always happy to interact with all sorts, he has never been the type of person to make anyone feel inferior. He is very outgoing and able to start a conversation with anyone, especially pretty ladies. As a child I found it fascinating that these ladies thought he was so charming. I always thought he looked like Robert Redford, and I guessed maybe they did too. I've always known my Dad was charismatic, with his goofy, fun loving quips, but didn't quite get the attraction. He was just my Dad. However as a young adult, to see him lecture on the subject he loves, revealed a new side of him to me. I was truly awed. I never realized he was 'a genius' until he won the MacArthur award. Around the same time frame, I heard him referred to as the Godfather of genetic algorithms and then I realized he had a vision that nobody else had come up with. I realized he was a world renowned scientist and an inventor!

To this day he is my rock, sounding board, and cheer-leader. Through the years when I've been in a tough spot he has always been there for me, to bail me out and give sound advice of how to deal with the circumstances. Although I have not always agreed with his advice, in retrospect, I actually can't think of a time when I didn't change my mind eventually and see the wisdom of his thinking. I am so lucky to have John Holland as my parent, my mentor, and my friend.

By Gretchen Sleamon — also known as the eldest daughter.

21

5   Alison Butler

Last Thanksgiving my nine year old daughter, Sydney, had a school assignment to interview a family member from a previous generation. My dad was visiting us in Washington, D.C. and thus became the lucky interviewee. The resulting interview was sweet and concise. Sydney ended the interview, as any descendent of John Holland's would, with a moment of humor.

Now, as I sit down to write about my dad, I am dumb struck. I don't have a particular event or interaction to write about. I have a lifetime. So I offer a few random memories, a slide show of John Holland, my dad.

—One of my earliest memories is of my father reading to my sister and I. Many people will remember their parents reading to them — Winnie the Pooh, Mother Goose, possibly something a little more "out there", such as James and the Giant Peach. My dad read my sister and I The Lord of the Rings. Not JUST The Lord of the Rings, but The Lord of the Rings with such feeling and passion that I felt as though the Black Riders could hear me breathing.

—Night Walks. They didn't happen so very often, but when they did they were memorable. Sometimes when the weather was good and the mosquito population low, my dad would take us for night walks on our property. We set out in search of nocturnal wildlife and to listen for owls. I must admit I was never very comfortable during these walks. Humans are well trained to be nervous about what they cannot see. I, however, did trust that my father would not let me get eaten by the ogres that lurked in the darkness. Ogres be damned!

—I have a particularly vivid image of a wonderful sketch my dad drew for me. He had watched a deer approach the edge of our pond and swim across. This was a rare and special occurrence that I had missed. So he took out pencil and paper to recapture the quiet moment — the deer, the ripples across the pond, the old fallen oak tree that extended out into the water — all reappeared. This was so much better than a photograph because the image came from my father's hand.

—Sunday Breakfast. After my parents divorce my dad would take my sister and I out each Sunday for a lovely breakfast. As an Eight year old I found it fascinating and delightful that the pretty waitresses were so attentive. Was it our stunning good behavior? Our spectacular choice of fresh Danish pastry rather than Frosted Flakes? Later, as an adult, I realized it's not a bad thing to have a rather handsome and scintillating father.

—During a visit to Hawaii when I was nine, my dad discovered a tiny gecko caught in the thick web of a garden spider. He spent 20 minutes extracting the gecko from the web and the web from the gecko. One of many memories I have of my father's true compassion for wildlife and the little guy that needs a hand.

—Driving instructor. A slightly less pleasant but more humorous memory is that of my father teaching me to drive a car. Something to keep in mind is, when I was age 15, most cars were standard transmission (at least in my family). As we started our session, with my dad seated calmly in the passenger seat, everything was going well. Then we arrived at the steepest hill in Ann Arbor. As I chugged up the hill, much to my horror, the light (to spite me) turned red. I had to stop mid-hill. I tried to juggle the use of clutch, gear shift and gas peddle but rolled back and stalled, then rolled back and stalled again, then rolled back and stalled a third time. Tensions rose, exclamations were made. Let me just say it was a good thing that my dad had a fallback career, because driving instructor was not his calling.

—Finally there is the junk drawer purge. During a recent stay at my dad's house I became frustrated while searching fruitlessly in his 'junk drawer' (the drawer where one tosses odds and ends that don't really have another logical resting place) for a measuring tape. Finally I decided I would clean out the drawer and relieve it of some of its more useless contents. These items included, but were not limited to, ancient batteries, five or six partially used rolls of electrical tape and several amputated extension cords. Toward the back of the drawer I unearthed an old, yellowed envelope containing an equally yellowed plastic sandwich bag. The bag contained a dingy, greenish, dried plant-like substance. A smile curled at the edges of my mouth, I opened the bag to sniff. The contents were so old that the normally pungent odor I expected to greet my nose was completely absent. The only scent present was that of old plastic bag. Later, with great pride, I showed my dad the expertly organized 'junk drawer'. When I relayed the story of the antique substance I'd discovered, I noticed a wee twinkle in his eye. When I told him I sent it swirling down the pipes of the commode he looked slightly wistful and walked away. I swear I heard him mutter that my generation just didn't "get" his generation!

Throughout this slide show there is a common thread, a soundtrack playing quietly in the background. I'll risk sounding saccharine and reveal that the thread is the great fondness, humor and pride I feel for my dad. There is also an understanding of where many of my best (and not best) attributes come from.

To end my slide show, I loosely quote the conclusion of Sydney's school interview. My father is a highly respected scientist in the academic world, but to me he's also my dad "who loves to tell bad jokes, eat bacon and take naps."

6   Manja Holland

"My Poppa!"

Of bugs and mushrooms

"Every kid has a bug period . . . I never grew out of mine."
– E.O. Wilson

E.O. Wilson's quote from a recent NPR Science Friday interview (June 21, 2013) also fits my dad, and the sentiment captures one of the things I love most about him. No, my dad has not spent his life studying insects, but his life-long passion and wonder for all things nature is an inspiration for me and was no doubt one of the key influences in my own interest in science and the natural world. For my dad, the word "bug" in the E.O. Wilson quote could be replaced by nearly any other science or nature-related object or topic (and many others). When asked several years ago whether there is anything that does NOT interest him, my dad paused to think. He then replied with a smile and said, "professional football." While I am sure there are probably at least a few other things that do not capture his interest, that list is exceptionally short. Professional football was not even meant as a stand in for all sports. My dad has been a long-time University of Michigan hockey fan and has attended games for many years.

On a recent trip to our family's vacation home in the Upper Peninsula of Michigan, my dad found several clusters of mushrooms near the house. He was intrigued both because we rarely find mushrooms growing on the sand dunes near our house and also because he wasn't able to identify them. He collected one of the mushrooms and placed it on a white piece of paper for a couple of days to look at the spore pattern. This is something I recall him doing numerous times during my childhood with mushrooms from the woods behind our house. He sent an email message inquiring about the mushroom and requesting that we bring mushroom identification guide along in a few days when we were to be joining him at the house. When we arrived, he was excited to show us the mushroom and its unusual spore pattern. My dad is neither a trained naturalist nor a biologist, but this is the level of enthusiasm and curiosity he demonstrates for the natural world and many, many other things.

King Arthur (The Fish)

My dad and I share a great compassion for animals, and I attribute much of my interest and compassion for animals to him. When I was in college I left my pet fish, King Arthur, in my dad's care for a couple of weeks at our U.P. house. The fish happened to die under my dad's watch, through no fault of his. I learned later that he decided to give King Arthur a proper burial that day, kayaking him out into Lake Michigan in the midst of a thunderstorm. Perhaps it is unusual for adults to be saddened by the death of a fish — many people certainly wouldn't have given it much thought. To me, my dad's gesture demonstrated our mutual respect for all living creatures, an ethic that is of great importance to us both.

The Key Bird

What does the Key Bird say as it flies over the Antarctic?
Key-Key-Key-Reist (Christ) it's cold out here!

My dad is the king of the awful dad/grandfather jokes. This is probably one of his favorites — a classic. The best part of my dad's jokes is his own laughter after them. He often cracks himself up so much that he can barely breathe, sometimes even a few tears streaming down his face. My dad has a great, highly contagious laugh. This oxygen-depriving, tear-invoking laugh and laughing at one's own jokes both appear to be genetic. I have inherited these traits, and I believe my two and a half year-old son, Nels, has inherited them too. I am very proud!

Buzzer Bee!

"Here comes the Buzzer Bee! Buzz buzz buzz!"

My dad pretends his finger is a bumblebee approaching Nels' nose. Nels squeals in delight. Nels then returns the favor. Nels loves his poppa, and as a demonstration of this says repeatedly and very emphatically, "It's MY Poppa!!" He loves to see and learn about any creatures or other things my dad shows him. Very few people are fortunate enough to possess or maintain the level of wonder and curiosity that my dad has maintained throughout his life. I believe this is one of the things Nels loves most about my dad, too. It is this endless curiosity that is part of what makes my dad a great scientist, but it is also a key part of what makes him such a beloved grandfather to Nels. It is this tremendous sense of wonder and curiosity, and great passion and compassion for living things that I hope my son will also carry with him throughout his life.

John with his grandson, Nels — photo from Manja Holland

7   Michael Arbib

Long Ago in Ann Arbor, Michigan
La Jolla, September 2013

Von Neumann's Self-Reproducing Automata

Even though my specialty then was Pure Mathematics, during my last years as an undergraduate at Sydney University I became keenly interested in cybernetics and automata theory and was thus intrigued to read in the *Sunday Sun* an article reprinted from *Newsweek* on John von Neumann's theory of self-reproducing automata. An outline of the theory had been published in 1951, but a large manuscript had been left unpublished at his death. The article stated that this manuscript was then (1960) being prepared for publication by a group at the University of Michigan, but gave no further details. So I wrote a letter of enquiry addressed to "The Group Working on von Neumann's Theory of Automata, University of Michigan." In those days of air mail, long before email, one might expect a reply from the other side of the world in about 2 weeks, but months went by without a response. And then an envelope arrived, with a long letter from John Holland. He apologized for the delay in replying. The letter had been erroneously sent to the Department of Automotive Engineering (!) but had in due course made it to the Logic of Computers Group. (I remain impressed by the unknown expert on automobility who knew where Michiganders studied automata.) The von Neumann manuscript was being edited by Arthur Burks (whom I would later learn was not only a pioneer of computer design but also an eminent scholar of the American pragmatist Charles Sanders Peirce) and would be ready for publication in 6 months. After several years of non-publication, 6 months became known as "the von Neumann constant" — the book was finally published in 1966. Whether or not I learned this from John's letter, I found that other members of the group — Copi, Elgot and Wright — had published a paper on "Realization of events by logical nets" which was to be a focus of my first academic publication. I also learned that John was active in studying cycles in logical nets and iterative circuit computers. In this way, John and his colleagues in Ann Arbor formed an important early influence in my formation as an automata theorist.

Self-Organizing Systems

As a graduate student at MIT (1961-63), I spent much of my research time (other than that for my Ph.D.) with the group of Warren McCulloch whose seminal 1943 paper with Walter Pitts, "A logical calculus of the ideas immanent in nervous activity" forms, with Turing's 1936 paper, "On computable numbers, with an application to the *Entscheidungsproblem*," the basis for automata theory, and in 1962 I was given the honor of presenting a paper "Neurological Models and Integrative Processes" written by McCulloch, Jack Cowan and myself, at the conference *Self-Organizing Systems 1962* held that May in Chicago. The most notable line of the paper was McCulloch's aphorism "He who wants a sweetheart in the Spring would not be wise to wait for an amoeba to evolve her." (An early statement of the importance of getting the right initial conditions for an evolutionary algorithm!) More important for our present purposes, though, is that another talk, "Concerning Efficient Adaptive Systems," was given by one John H. Holland and thus introduced me both to him and to his interest in adaptation.

Ann Arbor, at Last

Following submission of my Ph.D. thesis in 1963, I set off in an old Buick (replaced by an old Ford after rolling the Buick in Nebraska) to tour the United States, alternating visits to universities with days of unalloyed tourism. Fairly early in the tour, I came to Ann Arbor and met John again and became fully aware of the bubbling energy and keen intelligence he packed into that short frame. I also met Art Burks (still working on that von Neumann book), Jim Thatcher (just completing a Ph.D. thesis on self-reproducing automata) and Richard Laing (who had an early interest in embodied computation). But my strongest memory of the visit was attending the Michigan-Michigan State football game, where Michigan's then Governor Romney (Mitt's dad) diplomatically sat on one side of the stadium in the first half, then switched to the other.

Fragments

A few fragmentary memories of other visits to Ann Arbor.

Visiting, relatively newly married, with my wife Prue what I think was then John's rather new house on the rural outskirts of Ann Arbor and meeting John and Mary Ann and their two daughters, Gretchen and Alison. A beautiful house. I can only vaguely picture it now. Was it by a river?

On another visit, John introduced me to his work on genetic algorithms. I was impressed by the breadth of applications, from biology all the way to economics. I was honored to be asked to write a blurb for the University of Michigan Press when John's book on the topic was about to be published. It took some years for the book to come to fruition, so I was able to praise "this long awaited book." However, I had no real appreciation at that time of how significant John's innovations were, spawning in later years a whole new field of computer science with so many important applications.

On another visit, I was discussing with John and his wife Maurita my puzzlement that after the fall of the Roman empire, people who had enjoyed the benefits of fresh water allowed the aqueducts to fall into disrepair. Maurita, a librarian, responded by finding in the UM library an old sepia-toned volume on the construction of the aqueducts and the bureaucracy that kept them running. Presumably, disorganization, the damage of war, and the falling away of masonry skills all contributed to the ensuing decline.

Regrets
It has been far too many years since last I saw John. As the years go by, the foci of scientific lives change, and conferences that drew us together now keep us apart. But I cherish John and his enthusiasm and the ideas we have shared and his generous response to an undergraduate in far-off Australia more than half a century ago.

8   W. Brian Arthur

John Holland and the World of Economics[1]

John Henry Holland entered my life in an indirect way, one that came out of a single incident. In April 1987 I was walking towards my office in Stanford when a helmeted Kenneth Arrow swung round me on his bicycle and stopped. Arrow was putting together a group of economic theorists in September to exchange ideas with a group of physicists that his counterpart, physicist Philip Anderson, would propose. The venue would be a small institute in the Rockies just starting up. It was in Santa Fe. Would I like to come? I said yes immediately without being sure of what I was committing myself to. Somehow the idea looked promising.

The conference in Santa Fe a few months later turned out to be a more heavyweight affair than I'd imagined. Among the ten or so economists Arrow chose were Larry Summers, Tom Sargent, Jose Sheinkman, and William (Buz) Brock. Among the ten or so "physicists" Phil Anderson chose were John Holland, David Ruelle, Stuart Kauffman, and David Pines. The meeting was held in the chapel of a convent the new institute was renting and there was nothing rushed about it. A participant would talk in the morning and we would discuss, another participant would talk in the afternoon and again we would discuss. This sounds like a relaxed tempo but I found it intellectually intense. I was sharing a house with Holland at the time — I found out later that we were the Institute's first visiting fellows — and discussions between us continued well into the evening.

We were all of us learning together, not just solutions to problems in the others' discipline, but about what each discipline saw as a problem, and how it thought about these, and what mindset it brought to bear on these problems. Questions not normally raised within economics were raised — why do you guys cling onto perfect rationality? Why do you assume so much linearity? And questions were asked of physics too. Why is a problem "solved," say in spin

---

[1]   These pages have been excerpted from the preface to my forthcoming book, *Complexity and the Economy*, Oxford University Press, N. York, 2014.

glasses, when it has not settled to a steady state? Chaos theory and nonlinear dynamics were discussed in both economics and physics. Modeling of positive feedbacks and of interactions, again in both disciplines, was discussed. People would meet at night in twos and threes to talk over ideas and problems.

The meeting was exhilarating — and exhausting. Nothing had quite been solved by the end of the ten days, yet the physics side was left with a respect for the sheer complicatedness of the economy — the elements in the economy (people), unlike the ions in a lattice, could decide what to do next not just based on the current situation of themselves and other elements, but on what they thought those other elements might do given what *they* might do. And the economists were left with a feeling for modern physics, for its interactions and nonlinearities, its multiple possible end states, its lack of predictability — indeed for its complicatedness.

Word began to leak out after the conference that something interesting had happened at Santa Fe and the new institute's Science Board decided it would follow the conference up by initiating a long-term research program on the Economy as an Evolving Complex System. Holland and I were asked to come to Santa Fe the following year to head this. I had a sabbatical coming from Stanford and accepted, John found it harder to get away from Michigan and declined. So I found myself heading up the Santa Fe Institute's first research program; it would start in August the following year, 1988.

My immediate problem of course, working from Stanford, was to put together a team of first-rate people for the new program and to decide its direction. Some people I already knew from the conference. John promised to come for a couple of months, and the physicist Richard Palmer for much longer than that. Stuart Kauffman would be in residence. From my own network I was able to bring in David Lane and Yuri Ermoliev, both excellent probability theorists. Arrow and Anderson helped greatly. Where I found it hard to cajole people to join in, Arrow or Anderson, both Nobel Prize winners, could simply lift the phone and quickly get people to join in. As to direction I was less sure. Early on, the physicist Murray Gell-Mann suggested to me that we come up with a manifesto for doing economics differently. I didn't quite have the confidence for that, in fact I didn't yet know what topics we would go after. I had done quite a bit of work already on complexity and the economy, but now we had a much broader reach in what topics we might research. From the conference it was assumed that chaos theory would be central, but the idea somehow didn't appeal to me. Vaguely I thought that we should look at increasing returns problems which I was more than familiar with, at how some of the physics methods could be transferred into economics, and at nonlinear dynamics in the economy. Also we might be able to do something interesting with computation in economics.

When the program opened finally in 1988 we discussed direction further, still groping for a way forward. I phoned Ken Arrow from Santa Fe and asked for his advice and Phil Anderson's. They got in touch with the funder of the program, John Reed of Citibank, and the word came back: Do what you want providing it deals with the foundations of economics, and is not conventional. For me and the others on the team, this directive seemed like a dream. We had carte blanche to do what we wanted, and at Santa Fe we wouldn't have colleagues from the discipline looking at us and asking why we were doing things differently.

In fact, outside our small team the few colleagues we did have were from physics or theoretical biology. Stuart Kauffman was one, and we immediately included him in the program. There was little else in the way of researchers the new institute could offer. It was in its earliest days and was all but unknown, an experiment, a small startup in the Rockies set up to have no students, no classes, no departments, and no disciplines — no discipline, the wags said.

We had discussions, mainly in the convent's kitchen, and I remember in an early one Kauffman said, Why do you guys do everything at equilibrium? What would it be like to do economics out of equilibrium? Like all economists I had thought about that, but not seriously. In fact the question took me aback, and it did so with the other economists. I had no good answer. It fell into the category of questions such as what would physics be like if the gravitational force were reversed, something that seemed perfectly thinkable as a thought experiment, but strange. And yet Kauffman's question stuck. We retained the question but we were still looking for a direction ahead.

One of the directions that interested me was still half formed. It had come out of the conference the previous year. In an after-lunch talk the first day of that conference, John Holland had described his work on classifier systems, basically systems that are concatenations of condition-action rules. One rule might say that *if* the system's environment fulfills condition A, *then* execute action R. Another might say, *if* it fulfills condition D, execute action T. A third might say that *if* A is true, *and* R-being-executed is not true, *then* execute action Z. And so on. The actions taken would change the environment, the overall state of the system. In this way you could string such if-then rules together to get a system to "recognize" its environment and execute actions appropriately, much as an E-Coli bacterium "recognizes" a glucose gradient in its environment and swims in an appropriate direction. Moreover, you could allow the system to start with not-so-good rules and replace these with better ones it discovered over time. The system could learn and evolve.

As Holland talked about this I found myself deeply excited, and I checked the room to see if other economists were similarly taken with these ideas. There was no evidence, in fact one of them was taking a post-lunch nap. A feeling grew in me that somehow, in some way, this was an answer and all we had to do was find the question. Somehow Holland was describing a method whereby "intelligence" or appropriate action could automatically evolve within systems. I quizzed John later about his ideas. We were sharing a house in Santa Fe for 2 months at that time in 1987, but in several conversations neither of us could work out what these ideas might directly have to do with economics.

I had gone back to Stanford, where I was teaching a course in economic development. It occurred to me, gradually at first, that John and I could design a primitive artificial economy that would execute on my computer, and use his learning system to generate increasing sophisticated action rules that would build on each other and thus emulate how an economy bootstraps its way up from raw simplicity to modern complication. In my mind I pictured this miniature economy with its little agents as sitting in a computer in the corner of my office. I would hit the return button to start and come back a few hours later to peer in and say, oh look, they are trading sheep fleeces for obsidian. A day later as the computation ran, I would look again and see that a currency had evolved for trading, and with it some primitive banking. Still later, joint stock companies would emerge. Later still, we would see central banking, and labor unions with workers occasionally striking, and insurance companies, and a few days later, options trading. The idea was ambitious and I told Holland about it over the phone. He was interested, but neither he nor I could see how to get it to work.

That was still the status the following summer in June 1988 when Holland and I met again in Santa Fe shortly before the program was to start. I was keen to have some form of this self-evolving economy to work with. Over lunch at a restaurant called Babe's on Canyon Road, John asked how the idea was coming. I told him I found it difficult, but had a simpler idea that might be feasible. Instead of simulating the full development of an economy, we could simulate a stock market. The market would be completely stand alone. It would exist on a computer and would have little agents — computerized investors that would each be individual computer programs — who would buy and sell stock, try to spot trends, and even speculate. We could start with simple agents and allow them to get smart by using John's evolving condition-action rules, and we could study the results and compare these with real markets. John liked the idea.

We began in the fall, with the program now started, to build a computer-based model of the stock market. Our "investors," we had decided, would be individual computer programs that could react and evolve within a computer that sat on my desk. That much was clear, but we had little success in reducing the market to a set of condition-action rules, despite a number of attempts. The model was too ad-hoc I thought — it wasn't clean. Tom Sargent happened to be visiting from Stanford and he suggested that we simply use Robert Lucas's classic 1978 model of the stock market as a basis for what we were doing. This worked. It was both clean and doable. Lucas's model of course was mathematical, it was expressed in equations. For ease of analysis, his investors had been identical; they responded to market signals all in the same way and on average correctly, and Lucas had managed to show mathematically how a stock's price over time would vary with its recent sequence of earnings.

Our investors, by contrast, would potentially differ in their ideas of the market and they would have to learn what worked in the market and what didn't. We could use John's methods to do this. The artificial investors would develop their own condition/forecast rules (e.g. if prices have risen in the last 3 periods *and* volume is down more than 10%/ *then* forecast tomorrow's price will be 1.35% higher). We would also allow our investors to have several such rules that might apply — multiple hypotheses — and at any time they would act on the one that had proved recently most accurate of these. Rules or hypotheses would of course differ from investor to investor; they would start off chosen randomly and be jettisoned if useless or recombined to generate potential new rules if successful. Our investors might start off not very intelligently, but over time they would discover what worked and get smarter. And of course this would change the market; they might have to keep adjusting and discovering indefinitely.

We programmed the initial version in Basic on a Macintosh with physicist Richard Palmer doing the coding. (I had found to my surprise that John wasn't much of a programmer, he was at heart a theorist.) Initially our effort was to get the system to work, to get our artificial investors to bid and offer on the basis of their current understandings of the market and to get the market to clear properly, but when all this worked we saw little at first sight that was different from the standard economic outcome. But then looking more closely, we noticed the emergence of real market phenomena: small bubbles and crashes were present, as were correlations in prices and volume, and periods of high volatility followed by periods of quiescence. Our artificial market was showing real-world phenomena that standard economics with its insistence on identical agents using rational expectations could not show.

I found it exciting that we could reproduce real phenomena that the standard theory could not. Holland and Palmer and I were aware at the time that we were doing something different. We were simulating a market in which individual behavior competed and evolved in an "ecology" these behaviors mutually created. This was something that couldn't easily be done by standard equation-based methods — if forecasting rules were triggered by specific conditions and if they differed from investor to investor their implications would be too complicated to study. And it differed from other computerized rule-based models that had begun to appear from about 1986 onward. Their rules were few and were fixed — laid down in advance — and tested in competition with each other. Our rules could change, mutate, and indeed "get smart." We had a definite feeling that the computer would free us from the simplifications of standard models or standard rule-based systems. Yet we did not think of our model as computer simulation of the market. We saw it as a lab experiment where we could set up a base case and systematically make small changes to explore their consequences.

We didn't quite have a name for this sort of work — at one stage we called it element-based modeling, as opposed to equation-based modeling. About three years later, in 1991, John Holland and John Miller wrote a paper about modeling with "artificial adaptive agents."[2] Within the economics community this label morphed into "agent-based modeling" and the name stuck.

---

[2] J.H. Holland and J.H. Miller, "Artificial Adaptive Agents in Economic Theory," *Amer. Econ. Assoc. Papers and Proceedings*, 81, 2, 365-70, 1991.

We took up other problems that first year of the Economics Program. Our idea was not to try to lay out a new general method for economics, as Samuelson and others had tried to do several decades before. Rather we would take known problems, the old chestnuts of economics, and redo them from our different perspective. John Rust and Richard Palmer were looking at the double auction market this way. David Lane and I were working on information contagion, an early version of social learning. And of course Holland, Palmer and I were looking at asset pricing — the stock market. I had thought that ideas of increasing returns and positive feedbacks would define the first years of the program. But they didn't. What really defined it more than anything was John's ideas of adaptation and learning.

Much has happened since those early days. The economics program at Santa Fe has gone on to define and develop an approach that has become Complexity Economics. Agent-based modeling has widened and blossomed from the early efforts of the late 1980s; it is now a standard tool in the social sciences. The Santa Fe Institute has become well known for its work in many other disciplines besides economics. And John himself has gone on to influence other fields. In 1992 he was awarded a MacArthur Fellowship, a very fitting tribute to a very creative man.

What in the end is John Holland's influence in economics? It would be a mistake to say that he just taught us techniques — evolutionary algorithms, agent-based approaches, classifier systems, inductive reasoning. He did, but he also taught us much more than that. John taught us that for the systems that interested us, economic ones, *non*equilibrium rather than equilibrium is the norm. In such systems, discovery and novelty never stop, they are ongoing and perpetual. He taught us that systems cumulate the results of their explorations and experience, and as they do these are reflected in changes in their structure, and in new structures built from existing structures. He taught us — certainly he taught me — to think algorithmically: this causes that, and that further causes either this event or some other one. John's universe is an unfolding of verbs: it evolves, it changes, and it never repeats.

John Holland has influenced economics in a very John-like way, casually and almost unintentionally. But the field and our thinking in it will never be the same.

9   Robert Axelrod

How the Genetic Algorithm Helped Explain Sexual Reproduction
Contribution to book in honor of John Holland
Sept. 2013

I have had the privilege of working with John Holland for almost forty years, including a decades-long participation in the BACH group. Perhaps John's greatest intellectual contribution to science is his development and analysis of his Genetic Algorithm (Holland 1975; Riolo 1992). The Genetic Algorithm is a way of solving a wide range of very hard problems using artificial intelligence techniques inspired by the genetic mechanisms of biological evolution. I use this opportunity to recount one of the most usual and productive uses of the Genetic Algorithm, namely its ability to pay back its debt to evolutionary biology by fostering advances in that very field.

One day about 1987, one of the world's leading evolutionary biologists and a former member of the BACH group, William Hamilton, told me about a truly amazing theory he was developing.[3] The theory proposed an answer to one of biology's largest unresolved puzzles: why have most large animals and plants evolved to reproduce sexually? The reason this is such a puzzle is that sexual reproduction has a huge cost: only half the population has offspring. What might be the advantage of sexual reproduction that is so great that it can overcome this two-fold cost compared to asexual reproduction?

There was already a serious contender whose leading advocate was the Russian geneticist Alexei Kondrashov. Kondrashov's explanation was based on the possibility that sexual reproduction might be helpful for bearing the cumulative burden of many generations of deleterious mutations. Bill's theory was completely different. Put simply, he

---

[3]   Parts of this chapter are adapted from "Agent-Based Modeling as a Bridge Between Disciplines," in Leigh Tesfatsion and Kenneth Judd (eds.), *Handbook of Computational Economics, Vol. 2: Agent-Based Computational Economics* (New York: North-Holland), 2006, pp. 1565-84. Reprinted by permission.

thought of sexual reproduction as an adaptation to resist parasites.[4] This struck me as a totally bizarre, but intriguing idea.[5]

Bill explained to me that there was a serious problem with convincing others that his theory could, in fact, account for the two-for-one burden of sexual reproduction. The problem was that the equations that described the process were totally intractable when the genetic markers had more than two or three loci. Yet, the whole idea relies on there being many loci so that it would not be trivial for the parasites to match them. When I heard this, I responded to Bill with something like, "No problem. I know a method developed by John Holland to simulate the evolution of populations with a lot of genetic markers. The method is called the Genetic Algorithm, and I've already used it to simulate a population of individuals each of whom has seventy genes."[6]

I explained to Bill that John Holland is a computer scientist who had been inspired by the success of biological evolution in finding "solutions" to difficult problems by means of competition among an evolving population of agents.[7] Based on the evolutionary analogue, including the possibility for sexual reproduction, John developed the Genetic Algorithm as an artificial intelligence technique. I could simply turn this technique around and help Bill simulate biological evolution, with or without sex. Since Bill was used to thinking in terms of heterogeneous populations of autonomous individuals, he readily grasped the idea of agent-based modeling. He also grasped without difficulty that an agent-based simulation was capable of demonstrating that certain assumptions are sufficient to generate

---

[4] Bill liked this formulation of mine, and we used it as the title of our paper.

[5] Bill's reasoning was that parasites are ubiquitous, and their short life spans give them the advantage of being able to adapt quickly to an ever-changing host population. If the host population reproduced *asexually*, a line of parasites that had evolved to mimic the genetic markers on the cells of one host would automatically be well adapted to mimic the genetic markers of its offspring. On the other hand, if the hosts reproduced sexually, their offspring would not be virtual carbon copies of either of their parents, and thus would not be as vulnerable to a line of parasites that had become adapted to match the genetic loci of one parent or the other.

[6] I had previously used this evolutionary technique to avoid having to run new tournaments indefinitely. See Axelrod (1987).

[7] See Holland (1975 and 1992), and Riolo (1992).

certain results, even if the same results could not be proven mathematically.

So, working with a computer science graduate student, Reiko Tanese, we built an agent-based model with two co-evolving populations: hosts with long life spans, and parasites with short life spans. If a parasite interacted with a host of similar marker genes, it killed the host and reproduced. In the simulation, the parasite population would tend to evolve to concentrate in the region of the "genetic space" where there were many hosts. Thus, successful hosts tended to suffer from increasing numbers of deadly parasites, reducing the numbers of those hosts. Meanwhile, other types of hosts with very different genetic markers might thrive. Then the process would repeat itself as the population of parasites tracked the ever-changing population of hosts. The system would always be out of equilibrium.[8]

Bill was pleased with the results of our agent-based simulations. He felt that

> the notion I had started with, that even against sex's full halving inefficiency the problem could be solved by looking at the need of a population to manoeuver against its many rapidly evolving parasites, with these differentiating resistance tendencies at many host loci (the more the better), had been vindicated. (p. 561)[9]

> Our model had achieved results that others had stated impossible with the tools we were allowing ourselves. Many of the dragons that had oppressed individual-advantage models in the past seemed to us to be slain. ...[O]ur explicit modeling of a large number of loci in a Red Queen situation[10] certainly was [new] and the increase of stability of sex that came with the growth of numbers of loci made the most dramatic feature in our results. (p. 602)

---

[8] Agent-based models are convenient for studying out-of-equilibrium dynamics. Real economies may be perpetually out of equilibrium, for instance if there is continual innovation (Nelson and Winter 1982). Systems far from equilibrium are notoriously difficult to analyze mathematically, and perhaps for that reason are often downplayed in neoclassical economics. Agent-based modeling allows the analysis systems that are far-from-equilibrium.

[9] All quotes are from Hamilton (2002)

[10] Bill is referring here to the character in *Alice Through the Looking Glass* who says, "It takes all the running you can do, to keep in the same place."

It is the paper that I regard as containing the second most important of all my contributions to evolution theory.[11] [It was...] the first model where sex proved itself able to beat any asex competitor immediately and under very widely plausible assumptions. (p. 560)

So that is how John Holland's Genetic Algorithm helped explain one the largest puzzles in evolutionary biology: why do almost all large animals and plants reproduce sexually, when the evolutionary cost of having only half the adult produce offspring is so high? Bill Hamilton was able to formulate a hypothesis that sexual reproduction is an adaptation to resist parasites. But until he learned about John's Genetic Algorithm, Bill was unable to demonstrate that his theory would actually account for the puzzling fact, even in principle. So John's idea of borrowing from evolutionary biology to develop a powerful tool for artificial intelligence was turned around to offer highly productive way to explore evolutionary biology itself.[12]

Bibliography

Axelrod, Robert, 1980a. "Effective Choice in the Prisoner's Dilemma," *Journal of Conflict Resolution*, 24, pp. 3-25. Included in revised form as part of Chapter 2 and Appendix A of Axelrod (1984).

_____ , 1980b. "More Effective Choice in the Prisoner's Dilemma," *Journal of Conflict Resolution*, 24, pp. 379-403. Included in revised form as part of Chapter 2 and Appendix A of Axelrod (1984).

_____ 1981. "Emergence of Cooperation Among Egoists," *American Political Science Review*, 75, pp. 306-318. Included in revised form as Chapter 3 and Appendix B of Axelrod (1984).

_____ 1984. *The Evolution of Cooperation* (NY: Basic Books).

_____ 1986. "An Evolutionary Approach to Norms," *American Political Science Review*, 80, pp. 1095-1111. Included with an introduction in Axelrod (1997c).

---

[11] Hamilton, Axelrod and Tanese (1990). Bill regarded his most important paper to be the one that presented his formal theory of inclusive fitness (Hamilton, 1964).

[12] For recent experimental verification of Hamilton's theory see Morran (2011).

_____ 1987. "The Evolution of Strategies in the Iterated Prisoner's Dilemma," in Lawrence Davis (ed.), *Genetic Algorithms and Simulated Annealing* (London: Pitman, and Los Altos, CA: Morgan Kaufman, 1987), pp. 32-41. Included with an introduction in Axelrod (1997c).

_____ 1997a. "Advancing the Art of Simulation in the Social Sciences," in Rosaria Conte, Rainer Hegselmann and Pietro Terna (eds.), *Simulating Social Phenomena* (Berlin: Springer), pp. 21-40. Included with an introduction in Axelrod (1997c). An updated version of this paper is forthcoming in a special issue on agent-based modeling in the *Japanese Journal for Management Information*, and is among the papers available at http://www-personal.umich.edu/~axe/

_____ 1997b. "The Dissemination of Culture: A Model with Local Convergence and Global Polarization," *Journal of Conflict Resolution*, 41, pp. 203-226.

_____ 1997c. The Complexity of Cooperation: Agent-Based Models of Competition and Collaboration, (Princeton, NJ: Princeton University Press). Some of the chapters in this book are available at http://www-personal.umich.edu/~axe/

_____ and Douglas Dion, 1988. "The Further Evolution of Cooperation," *Science*, 242, pp. 1385-1390.

_____ and Michael D. Cohen. 2000. *Harnessing Complexity: Organizational Implications of a Scientific Frontier* (NY: Free Press).

_____ and William D. Hamilton, 1981. "The Evolution of Cooperation," *Science*, 211, pp. 1390-1396. Reprinted in modified form as Chapter 5 of Axelrod (1984).

_____ and Tesfatsion, 2005. "A Guide for Newcomers to Agent-Based Modeling," Kenneth L. Judd and Leigh Tesfatsion (eds.), *Handbook of Computational Economics II: Agent-Based Computational Economics*. (New York: North-Holland), 2006.

Axtell, Robert, Robert Axelrod, Joshua Epstein, and Michael D. Cohen, 1996. "Aligning Simulation Models: A Case Study and Results," *Computational and Mathematical Organization Theory*, 1, pp. 123-141.

Dawkins, Richard, 1976. *The Selfish Gene*. (Oxford: Oxford University Press). New Edition 1989.

Hamilton, William D. 1964. "The genetical evolution of social behaviour. I and II," *Journal of Theoretical Biology*, 7, pp. 1-52.

_____ , 2002. *Narrow Roads of Gene Land: The Collected Papers of W. D. Hamilton, volume 2, The Evolution of Sex*. (Oxford, Oxford U Press), pp. 117-132, 561-566, and 601-615.

_____ , Robert Axelrod and Reiko Tanese. 1990. "Sexual Reproduction as an Adaptation to Resist Parasites," *Proceedings of the National Academy of Sciences (USA)*, 87, pp. 3566-3573.

Hoffmann, Robert, 2000, "Twenty Years on: The Evolution of Cooperation Revisited," *Journal of Artificial Societies and Social Simulation* vol. 3, no. 2. Available at <http://www.soc.surrey.ac.uk/JASSS/3/2/forum/1.html>

Holland, John H, 1975. *Adaptation in Natural and Artificial Systems: An Introductory Analysis With Applications to Biology, Control, and Artificial Intelligence*. Ann Arbor, MI: University of Michigan Press. Reissued Cambridge, Mass.: MIT Press, 1992.

_____ , 1992. "Genetic Algorithms," *Scientific American*, July, pp. 44ff.

Luce, R. Duncan and Howard Raiffa, 1957. Games and Decisions; Introduction and Critical Survey (New York, Wiley).

Maynard Smith, John, 1982. *Evolution and the Theory of Games* (Cambridge, UK: Cambridge University Press).

Miller, John H. 1998. "Active Nonlinear Tests (ANTs) of Complex Simulations Models," *Management Science*, 44, pp. 820-830.

Morran, Levi T., Olivia G. Schmidt, Ian A. Gelarden, Raymond C. Parrish, and Curtis M. Lively. 2011. "Running with the Red Queen: Host-Parasite Coevolution Selects for Biparental Sex." *Science*, 333, pp. 216-218.

Nelson, Richard R. and Sidney G. Winter, 1982. *An Evolutionary Theory of Economic Change* (Cambridge: Harvard University Press).

Riolo, Rick L., 1992. "Survival of the Fittest Bits," *Scientific American*, July, page 89ff.

Samuel, Arthur, 1959, "Some Studies in Machine Learning Using the Game of Checkers," *IBM Journal of Research and Development*, vol. 3, no 3, pp. 210-229.

Schelling, Thomas, 1978. *Micromotives and Macro_Behavior* (NY: Norton).

Taylor, Michael, 1976. *Anarchy and Cooperation* (New York: Wiley).

Trivers, Robert, 1971. The Evolution of Reciprocal Altruism. *Quarterly Journal of Biology*, 46, pp. 35-57.

_____ , 2000. "Obituary: William Donald Hamilton (1936-2000)," *Nature*, 404, pp. 828.

Turner, P. E. and L. Chao, 1999. "Prisoner's Dilemma in an RNA Virus," *Nature*, 398, pp. 367-368.

Wu, Jianzhong and Robert Axelrod, 1995. "How to Cope with Noise in the Iterated Prisoner's Dilemma," *Journal of Conflict* Resolution, 39, pp. 183-189. Reprinted with an introduction in Axelrod (1997c).

10   Robert L. Axtell

Holland, Emergence, and the 'Inside-Outside' Dichotomy

During my last semester of graduate school at Carnegie-Mellon, while busily writing my thesis, I attended the Santa Fe Institute's first and, I believe, only Winter School on Complex Systems, in January of 1992 in Tucson. At the time a Santa Fe-inspired research thrust known as 'artificial life' (ALife) was bearing initial fruit, mixing simple biological models with the microcomputers of that era to give rise to minimal, self-reproducing structures that mimicked life forms, either real or imagined. In the fall of that year, now as a Research Fellow at the Brookings Institution in Washington, D.C., I undertook work, together with Joshua Epstein, that would come to be importantly influenced by John Holland.

Josh and I were each familiar with the ALife work going on at SFI and in September of 1992 we sketched out a simple computational model of basic social processes, which we called 'Artificial Social Life'. Over the next month, in a windowless office on the 5th floor at Brookings, I wrote a few thousand lines of code that would become known as the Sugarscape model, after the name we gave the resource-strewn landscape on which Artificial Social Life took place. Right away the model began producing results that seemed quite interesting to us, which was more than a little surprising because our agents were following very simple and purely local rules. Initially, we tended to 'explain' our results by stating what was going on at the individual agent level, and then simply describing the kinds of patterns we saw at the aggregate level, pertaining either to groups of agents or to the whole population.

That November there was going to be a workshop at Michigan on evolutionary computation, artificial life, computational models of social processes, and associated topics, related to the long-running efforts of the BACH research group there, of which John Holland was a member, of course. At the time I did not know anything about BACH, although in later years Josh and I would visit regularly. Well, Josh contacted the organizers of the Michigan meeting and right away we began preparing our talk. Arriving in Ann Arbor I met the BACH principals for the first time. I also recall being highly impressed with the other faculty and students who took part in the workshop. Today, from my perspective of 20 years of research on complex systems and agent-based models, it is very clear that BACH did much to prepare the way for the coming wave of work on complex adaptive systems, particularly in the social sciences.

The talk we gave included animations of our model, made possible by a state-of-the-art LCD display that sat on top of the overhead projectors of the day, making clear to the audience the phenomena we had produced. We offered aggregate descriptions of the regularities the model was producing in addition to stating the behavioral rules the agents were following. I remember Bob Axelrod and the late Michael Cohen being enthusiastic about what we were doing, and asking many questions directly relevant to the implications of the model for the social sciences. It was also the case that John made a series of comments that I did not feel I even knew how to parse, let alone answer. These had to do with *emergence*.

John couched his queries in terms of the agent level of the model being qualitatively different from the aggregate level at which the phenomena were visible. His metaphor was the rain forest in which the individual level is constantly in flux, with some animals surviving while others perish, yet at the aggregate level rain forests are incredibly robust and resilient ecological systems. The Sugarscape model was illustrating some aspects of his metaphor, as the agents never settled down into anything like a fixed point (e.g., Nash) equilibrium, yet the macro-level did seem to reach something like a stationary state. But he didn't seem to like the fact that we always spoke of our results in purely reductionist terms, without attempting to talk about *how* the macro-level acquired its distinct properties, features that made it unlike the micro-level. Once our talk was over he came up to us and pressed us further, in gentle terms and with humble demeanor yet in a confident and knowing way. He mentioned a variety of work he felt was related, none of which, I remember vividly, had I ever heard of, which seemed strange since, as a newly minted Ph.D., I felt I had reasonable command of the research literature in the areas we were working in.

Only later would I learn that two distinct projects John was working on at the time were directly related to ours, and so he had clearly been thinking about the issues he raised a lot longer and more deeply than we had. Specifically, his ECHO model, with Terry Jones, was also about simple creatures (agents) migrating around a resource landscape. Their modeling goals were quite different from ours — they more focused on abstract ideas of tags and cell walls —

but clearly the models had much in common. At the same time John was also working with Brian Arthur on an agent-based financial market model, along with Blake LeBaron, Richard Palmer, and others (Palmer *et al.* 1994; Arthur *et al.* 1997). That model too had the 'rain forest property' of perpetual adaptation by the individual agents, as they tried to outmaneuver one another trading, while regularities in price and volume statistics arose that could be made to qualitatively resemble data, on financial markets in their case.

At any rate, in the wake of attending that meeting at Michigan I tried to dig out some of the references John had mentioned, and tried to learn more about emergence. The Brookings library, which takes up an entire floor of the building, had served me well in my first months there, being rich in social science content. But when it came to the philosophy of emergence or the use of the term in computer science and AI, that library had few holdings. Multiple trips to Foggy Bottom and the Gelman Library at George Washington University were only slightly more productive. So I decided to contact John directly, asking him to point me in the direction of some reading I might do on the subject of emergence, and to clarify what he felt would constitute a satisfactory explanation of what our model was doing. At this point neither of John's books from later in that decade (Holland 1995; 1998) were available for him to cite, each being quite germane to my query. What I received was very specific to my problem and extraordinarily detailed. Indeed, I remember being both surprised and grateful that John had taken the time to send me so much information, running to several pages, including background material and references, which addressed my questions very thoroughly and precisely.

I had come to agent modeling through an undergraduate engineering degree and graduate training in computer science, economics and game theory and, despite having sympathies for notions of 'the whole being greater than the sum of its parts,' I had no formal bases on which to think through what this meant, either in general or in the case of the Sugarscape model. John's message to me couched ideas of emergence in terms of the 'inside-outside' dichotomy in artificial intelligence.

A particular concern of John at that time, and something he has continued to write about since, is the idea of a boundary and how it arises or, in biological terms, gets constructed. A system, whether biological or computational or social, that can create 'cell walls' and thereby generate both an inside and an outside, is capable of generating an identity, and doing so in hierarchical fashion. John emphasized to me the importance of each of these things —

boundary, identity, hierarchy. Often John and his collaborators used 'tags' to help agents distinguish one another, and these proved incredibly powerful in a variety of subsequent work we did at Brookings, notably in the culture and conflict portions of the Sugarscape model. Tags proved sufficient to generate conditional behavior, creating heterogeneous outcomes from behaviorally homogeneous agents, which led to further diversity at the agent level and endogenous inequality at the social level, all features of the real social world we sought to generate from the bottom up.

It is at this time that I began to see ideas surrounding emergence as being of central importance for what we were trying to do with agent systems. In subsequent discussions with Herb Simon it was clear he recognized the role SFI and the ALife communities were playing in the dissemination and popularization of ideas concerning emergence. He went on to add a new chapter in the 3rd edition of his *The Sciences of the Artificial* (Simon 1996) to deal with Santa Fe's contribution. At about that time Bob Axelrod was working on models in which new political actors emerged (Axelrod 1995), while his student, Lars-Erik Cederman, was trying to apply such models to actual geopolitical landscapes (Cederman 1997). Soon I began working on the emergence of firms (Axtell 1999). Having John's abstract, theoretical 'take' on what these models with emergent groups were actually doing was important for me, especially insofar as some mainstream economists, at Brookings and elsewhere, seemed to have a hard time situating what I was doing within the framework of conventional economics — wasn't I just building a computational model of Ronald Coase's transaction costs theory of the firm and, if so, what was the use of that?

Subsequently my colleague Epstein began expressing his discomfort with some of the ways emergence was being used in the growing complex systems literature (Epstein 1999). During discussions with him I would often return to John's framing of the main issues involving emergence, and while I do not think I convinced him, he and I did go on to use 'emergence' in the titles of papers (e.g., Axtell *et al.* 2001). Eventually, with statements of the importance of studying emergent phenomena appearing in high visibility places (e.g., Laughlin and Pines 2000), I would come to see that the multi-level character of social science, in which the individual agent level is conceptually distinct from the aggregate, macro-social level, places prime importance on ideas of emergence, and serves to differentiate the complex adaptive systems research program from the more conventional, reductionist program that was being pursued when I was in graduate school, and which continues to command a significant fraction of published pages in leading social science journals today.

In all this John Henry Holland was a pioneer. He led the way conceptually. He supported younger people professionally. He helped us think about things in clear, useful and novel ways. His impact on the complexity field cannot be overestimated, and his role in fomenting the agent computing vision among social scientists is second to none among computer scientists. When the definitive history of the complex systems field is someday undertaken, surely John's role will be central to the narrative.

References

Arthur, W. B., J. H. Holland, B. LeBaron, R. Palmer and P. Tayler (1997). Asset Pricing Under Endogenous Expectations in an Artificial Stock Market. *The Economy as an Evolving Complex System II*. W. B. Arthur, S. N. Durlauf and D. A. Lane. Reading, Mass., Addison-Wesley.

Axelrod, R. (1995). A Model of the Emergence of New Political Actors. *Many-Agent Simulation and Artificial Life*. E. Hillebrand and J. Stender. Washington, D.C., IOS Press.

Axtell, R. L. (1999). The Emergence of Firms in a Population of Agents: Local Increasing Returns, Unstable Nash Equilibria, and Power Law Size Distributions. *Working paper*. Santa Fe, N.M., Santa Fe Institute.

Axtell, R. L., J. M. Epstein and H. P. Young (2001). The Emergence of Classes in a Multi-Agent Bargaining Model. *Social Dynamics*. S. N. Durlauf and H. P. Young. Cambridge, Mass./Washington, D.C., MIT Press/Brookings Institution Press: 191-211.

Cederman, L.-E. (1997). *Emergent Actors and World Politics: How States and Nations Develop and Dissolve*. Princeton, N.J., Princeton University Press.

Epstein, J. M. (1999). "Agent-Based Computational Models and Generative Social Science." *Complexity* 4(5): 41-60.

Holland, J. H. (1995). *Hidden Order: How Adaptation Builds Complexity*. New York, N.Y., Perseus Press.

_____(1998). *Emergence: From Chaos to Order*. Reading, Mass., Perseus.

Laughlin, R. B. and D. Pines (2000). "The Theory of Everything." *Proc Natl Acad Sci USA* **97**(1): 28-31.

Palmer, R. G., W. B. Arthur, J. H. Holland, B. LeBaron and P. Tayler (1994). "Artificial economic life: A simple model of a stock market." *Physica D* **75**: 264-274.

Simon, H. A. (1996). *The Sciences of the Artificial*. Cambridge, Mass., MIT Press.

11   Ronda Butler-Villa

About John

When I was first asked to contribute to a book about John Holland, I declined because I was busy and felt others would write about John Holland's contribution to science. Why would I, a non-scientist, have anything to offer? Later I decided that John has taught me *about* science and scientific inquiry, and maybe that would be a useful addition to such a book, compiled by his friends. Here's some of what I've learned from him:

It's the road, not the destination. In 1988, John Holland emerged from the chapel-turned-conference-room. He was smiling widely. New to the Santa Fe Institute, I asked him, "What happened?" He kept smiling and said, "I don't know. It's very exciting." Naively I was wondering what discovery or solution or model they had created during the workshop. I was looking for a tangible result. John's enthusiasm reminded me that it was not so much about discovery as the excitement of scientific discourse, possible avenues of fruitful research, and thoughtful collaboration. His response helped me to understand what kind of place the Institute would be — a place to seek understanding of how things work — and his enthusiasm was contagious.

Agree on the big picture. In those early years he also helped me understand that researchers can agree on a concept without agreeing on the particulars of that concept. I had read Murray Gell-Mann's definition of "complex adaptive systems." Not long thereafter, I asked John for his definition. They were not the same, nor were they similar. When I pointed this out to John, he seemed unconcerned. Agreement, he explained, is possible regarding larger concepts without agreeing on a specific definition. Some people who understand this might use the example of "poverty"; we all agree it exists but not on what defines it. For me, it was like saying the word "dog," which everyone agrees they understand though each might visualize a different kind of dog when they hear the word. However, I did give up on the idea of a glossary of complex systems as every term might lead to three or more annotated definitions!

Enthusiasm can be unselfish. I have met many people who are excited about their own work and able to convey that excitement. I have met a few people who are also excited about others' work and able to convey *their* ideas with excitement and clarity. John is one of those people. (Erica Jen is another.) As I said, his enthusiasm is contagious. As Laura Ware puts it, he makes others feel good about themselves. He supports others, clarifies the nuggets of wisdom in their ideas, and compliments good ideas or intriguing lines of thought.

Leaders set a benchmark, sometimes unintentionally. After the Institute moved to its Hyde Park Road location, John was going to give an informal tutorial about complex systems research at the Institute. Both conference rooms were in use so he stood in a moderate-sized common space in front of a white board, with people walking by as they moved from one part of the building to another. While the sitting area usually allowed for 10 or so people, the group surrounding the whiteboard expanded, chairs were pulled up, and soon more than 30 people were listening — surprising, considering that John was competing with two other research events. After the tutorial, I overheard someone say how useful this had been and how we needed more tutorials like this one. John had set a benchmark for others to follow.

The honor of being first. For many years the Institute had a scientific book series. SFI created an Editorial Board to run the publications program, and these individuals determined which books, primarily proceedings volumes, would appear in the series. When our publisher, Addison Wesley Publishing, approached the Institute about doing a lecture series that would result in a different kind of book, the Editorial Board took this under consideration. This board wanted a good scientist, who could speak at a layperson's level, and who would provide thought-provoking lectures over three nights. The publisher wanted someone who had yet to write a trade book, someone who could write without a lot of equations, and someone with broad appeal. Interestingly, both institutions came up with a handful of possible candidates, but the one who stood out was John Holland. He was the first lecturer for the Institute's Stanislaw Ulam Lectures, giving three one-hour lectures over three nights plus a question and answer period after each lecture. His book "Hidden Order: How Adaptation Builds Complexity" was the result. With his clear writing style and use of accessible examples, John once again set a benchmark for others. Few have duplicated his success — an excellent lecture series and a well-received book.

Sometimes you need to dig deeper. I once mused to John how in the earlier years it was easier to pull out accessible ideas, concepts, and models to share with interested laypersons who came for SFI tours. I was finding that harder to do as the Institute matured. He flashed one of his smiles and explained that, early on, good ideas were like gems just lying on the ground, waiting to be picked up and examined. Those had all been discovered, so now we have to dig a little deeper. I too have learned to dig a little deeper.

Take responsibility. At an informal dinner party, I was surprised by John's candor. He was circumspect about his life and his shortcomings, in particular his divorce. He described how his choices had led to his current state. I never heard him blame another person for the things that had gone awry. He noted what happened and took responsibility for his actions or inaction. A noble man and a wonderful example.

I realize that memory isn't perfect and John may not realize all the times I carefully listened to what he, and others, had to say. I was lucky to have found the Santa Fe Institute and to work here, and I am thankful that it has allowed me to get to know some amazing people. One of them is John Holland.

12   Paul N. Courant

John Holland and Poker

John Holland is a poker player, and it will come as no surprise to those who know him and his other (non-poker) work that he is a very good poker player. As one would expect of a developer of algorithms and producer of both natural and artificial intelligence, John knows the probabilities, but there is much more to poker than knowing, for example, the odds of improving three-of-a-kind in two chances, conditional on what one has seen in the hand. In the words of a famous poker player who played with John for some forty years, one plays poker with cards, but against people.[13]

But that's not exactly right in John's case, for two reasons. One derives from the nature of the game that John has played in (with me as a participant for perhaps the last thirty years.) As a formal matter, there is nothing quite so zero-sum as poker, and there is no purer adversarial transaction than the showdown at the end of a hand, where, win, lose or draw the money that doesn't go to one player goes to the other(s), and vice versa. Yet the game as played by John is in an important sense collegial. We are playing with each other as well as against each other; it's no fun without the extended friendship and acquaintance of the others around the table. John is with the others as well as against them.

John has a remarkable skill at the poker table that also extends to the seminar room and the Santa Fe Institute. Whatever John is doing, he is up to something, only you don't know always know exactly what. In the seminar room, you get to wait and see, and there is likely an "aha" or two at the end of the discussion. At the poker table, the knowledge that John is up to something strikes fear into the hearts of men. What this comes down to is a remarkably intelligent use of intelligence on John's part. Everyone knows that John is very smart, and thus we infer that he knows what he is doing — check, call, raise or fold. There is a phrase in the poker literature called "limping in," and it refers to the behavior of players who don't force the action, but who

---

[13] See Peter O. Steiner, "Thursday Night Poker," revised edition, Ballantine Books, 2005, p.10.

play passively most of the time. John never limps in. Indeed, in contradistinction to his social demeanor, he is often downright pushy, and he has a penchant for raising on top of large bets by others, something that he clearly enjoys.

John consistently conveys to his fellow players that he is fearless — that he is willing (even delighted) to make moves that are risky. This technique leads to a certain inscrutability. When he doesn't make an aggressive play, is it because his position is weak or is it that his position is strong and he wants to convey that it's weak? The last time I raised him in this position, he raised me back and I folded. But maybe he was bluffing. But it would have cost a lot to find out. And it may cost a lot this time, too. So I don't know what to do. But if it's the same as last time it will cost me a lot, and if it's different I still won't know what to do. (One can get into a sort of infinite regress here, which I will abjure, but John's ability to get the others at the table to think this way is at the heart of his success as a poker player.)

There is a great deal of acting that goes along with all this. As everyone knows, John has a first-rate impish grin. Again, in seminar or discussion it's associated with some cool new idea or insight. And again, at the poker table, it makes the other players afraid. The impish grin conveys that he knows something that you don't — it could be about his hand, which is fine, but it also could be that he has figured out what someone else is up to, perhaps even what you are up to. So in a game where the key to success is to be enigmatic and unpredictable in one's own play, John goes one better and is enigmatic in his personal behavior. He may look puzzled, happy, knowing, troubled, amused, or even like he is listening to a faraway violin, and every one of these and more leads his friends around the table to be slightly more confused then they were before. Until the last card is dealt and the last bet is made, one does not know what John is up to.

For all of this skill, John is merely a very good player, rather than a great one. I think that this is a matter of choice on his part. He could take his considerable skills as an applied game theorist and producer of enigmatic expressions and win a good deal more money than he does. His saving grace (at least from my perspective as I figure out how I'm going to pay the mortgage) is that he likes the action. So he stays in some hands when the odds are not in his favor in the hope that one of those moments where he can really flummox everyone will emerge. If cards don't fall that way, or if one or more of the other players make bets that make it clear that in this instance the flummoxing move can't work, he will take the loss cheerfully. But if things go his way, he will not only win, and do so conveying the best of cheer, but also drive everyone else nuts.

It has been my great pleasure to know John in professional contexts as well as at the poker table, and to learn a great deal from him about models and modeling and how to think about all manner of hard problems. He of course loves figuring out new things and new ways of thinking, and has changed the way that many of us think about the world, while reminding us how much fun it is to think and understand in new ways. At the poker table one gets to see all of this in microcosm, with the additional pleasure (well, sometimes it's a pleasure) of watching him use his intelligence and understanding of the problem at hand in a context where it's relatively easy to keep score. And part of the score, for John, is the ability to drive everyone else nuts in a way that gives both him and everyone else great pleasure.

13   Tom Dawson

Administratively speaking

In 1966 I was a Marine returning from Vietnam, newly married, and starting undergraduate school. The plan was that Jinx, my wife, would teach high school while I attended school and worked part time. Jinx taught the first year while I learned to be a student. In the Fall of 1966, I was fortunate to find a part time position with the Logic of Computers Group, a research arm of the Department of Computer and Communication Sciences. The Directors of the Logic of Computers Group were Dr. Arthur Burkes and the Associate Director was Dr. John Holland.

In the summer of 1967 Jinx got pregnant, and under the rules of the day, she was forced into retirement which made me seek more full time work. When I reported my situation to John and Art, they simply increased my time to 90% at the Logic of Computers Group, also giving me a pay raise. In 1969 I became the administrative associate for the Department of Computer and Communication Sciences, splitting my time equally between the department and research and moving to full time status.

I completed my BBA in the Spring of 1969 and started law school at the Detroit College of Law, now Michigan State University Law School. My classes were downtown Detroit which required that I leave work early. I graduated in 1973, moved to Nebraska with my wife and two children to practice law. In 2006 I retired as a juvenile court judge and began to teach part time at the University of Nebraska and Nebraska Wesleyan University.

The foregoing would not have been as possible had it not been for the generosity of John and Art. Their willingness for me to study while on the job, as long as my duties did not suffer, allowed me to complete my education. Whenever I needed to miss work for class or a test in undergraduate school, no questions asked — school came first according to John and Art.

Later, in law school when I was full time at The University of Michigan I was permitted to leave my office and go the U of M law library to study. I was also allowed to take a class at The University of Michigan Law School during working hours, again with the caveat that my work for the department was done.

Due in part to the smallness of the department I got to know most of the students and even socialized with them, which made for a wonderful place to work. What made the job especially great was the relationship I had with John and Art, as well as the rest of the faculty and staff. I was treated as a friend and a colleague.

I did not grasp most of the science at the Logic of Computers Group, nor the study within the department, but to the extent I was willing to learn, John would teach me. John was a born teacher who could teach an unskilled, uninformed, neophyte, an ignorant law student enough to get excited about the research being conducted.

Eventually we were able to put the department and the research group together at the Frieze Building. This resulted in all the computers for the Logic of Computers Group to occupy the basement there. While the equipment would be considered prehistoric by today standards, it was state of the art for the day. When IBM developed a fast big printer, John took me to see this terrific piece of equipment — it was a chain printer and everyone was elated. I recall on one of our visits back to Ann Arbor we were visiting John at his home and he pointed to his desk computer and said "this is more powerful than the whole room of computers were at the Logic of Computers Group.

John and I have maintained contact over the years and I treasure that relationship.

My experience with the Logic of Computers Group and the Department of Computer and Communication Sciences was a very special chapter in my life and John played a significant role. He was an example of how people should be treated and how a leader should act. Many of these lessons I incorporated into my life.

John's place in the world of science is secure and I know those students that had the good fortune to have John as a mentor were indeed blessed. I have had the pleasant experience to have John as a boss, but more importantly as a friend. Thank you John for all you did for me and especially for your friendship.

Tom Dawson
Administrative Associate

14 Kenneth De Jong

University Professor
Professor of Computer Science
Associate Director of the Krasnow Institute for Advanced Study

As one gets older, the sense of which individuals have had a major impact on one's life sharpens. In my case there is no doubt that John Holland has had the strongest influence on my academic and professional life. As a graduate student in Computer Science at the University of Michigan in the late 1960s, I enrolled in Holland's Adaptive Systems course and was immediately infected by his innovative ideas, his boundless enthusiasm, and his interdisciplinary perspective. That was the first step in a life-long journey that included:

- A Ph.D. thesis that extended and hardened Holland's early adaptive systems Ideas into useful algorithms for dealing with difficult optimization problems in computer science and engineering.

- An academic life as a computer science professor continuing to extend these ideas to heuristic search and machine learning, as well as inspiring a new generation of Ph.D. students.

- A professional involvement in related activities, including being the founding editor-in-chief of the Evolutionary Computation journal, and a founding member of ACM SIGEVO.[14]

- A leadership role in the Krasnow Institute, a sister institute to the Santa Fe Institute.

My recollections of life as one of Holland's Ph.D. student in the early 1970s have somewhat faded over time, but in general characterized as pleasant, challenging and at times amusing. Here are a few recollections that give a sense of our interactions:

---

[14] SIGEVO: Special Interest Group on Genetic and Evolutionary Computation.

- Many lengthy discussions in which we struggled to find feasible computational interpretations of John's early ideas about genetic adaptive systems.

- Being continually amused by John's distrust in high-level programming languages and his strong preference for programming at the machine code level expressed in hexadecimal!

- Struggling to maintain discussion continuity with John in one of his offices that had windows that looked out on a variety of coed activities.

- Having delightful wide-ranging scientific discussions in the beautiful and informal setting of his hill-top home in the outskirts of Ann Arbor.

If we think of Holland's Ph.D. students as his academic children, now in his 80s John is the patriarch of an ever-expanding family tree of grandchildren and great-grandchildren. It's difficult to get an accurate count, but a rough estimate of at least ten of Holland's Ph.D. students active in academia, each producing 10-15 Ph.D. students of their own gives one a clear sense of the breadth of his impact. And, of course, a number of these grandchildren are now producing their own Ph.D. students.

In addition to his impact on and through specific individuals, Holland's innovative adaptive systems ideas continue to shape the fields of computer science and engineering in more general ways. His early ideas about genetic adaptive systems led to a class of genetic algorithms that continue to play an important role in the broader field of evolutionary computation, a field that has grown and matured in the past two decades to include a variety of respected journals, conferences with typical attendance figures of 400+, an ever increasing number of books, and an ACM special interest group SIGEVO.

His early adaptive systems ideas also included innovative ideas about machine learning, articulated as a "Learning Classifier System", that continues to play an important role in the reinforcement learning community. Both his work on classifier systems and genetic algorithms reflect the "bio-inspired" aspect of Holland's contributions. Today, that theme has matured into the field of "Natural Computation" that takes a broader perspective than Evolutionary Computation and is having a similar impact on computer science and engineering.

Perhaps the most under-appreciated aspect of Holland's contributions comes from his firm belief that innovation occurs on the boundaries of disciplines. This is perhaps best reflected in his early involvement in establishing the Santa Fe Institute and his continuing contributions to our understanding of complex adaptive systems. The important notions of emergence and agent-based modeling of complex systems have been fundamentally shaped by Holland's adaptive systems ideas. His influence continues today though his active participation in the Santa Fe Institute, the Krasnow Institute at George Mason University, and a variety of complex systems programs around the world.

It is difficult to know where to stop documenting Holland's impact on me personally as well as the broader fields of computer science and engineering. In my case, the most recent items relate to the fact that I am approaching the "normal" age of retirement and frequently asked what my plans are. I just smile and point to John as a role model who is still active in his 80s!

15   Alice Fulton

af89@cornell.edu

"Thirteen Ways Of Looking At John Holland"

I
On the five great lakes
the most buoyant cargo
was the mind of John Holland.

II
I was of three gladnesses
like a heart
in which there are three John Hollands.

III
John blushed in the February cold.
We made it a small part of the ceremony.

IV
A birdswarm and a fishflock
have no boss.
A birdswarm and a fishflock and John Holland
have no boss.

V
I do not know which to prefer
the innovations of the rainforest
or the innovations of the algorithm.
John making green tea
or John just being.

VI
Icicles fanged the study window
and the world turned bitter.
Still Lake Michigan
had no icefield in its middle.
Everyone said it
resembled John Holland.

VII
O Tolkien and Hesse,
why do you envision glass bead games,
the predictions of wizards?
Don't you see John Holland
on his hilltop unfurling hidden oracles?

VIII
I know small causes
can lead to large effects;
but I know, too,
an effect so large
its boundaries cannot be
defined. John Holland!

IX
When John Holland is disturbed
he is self-repairing
as a standing wave.

X
As John Holland is the mindchild
of a Lovelace and a Turing,
these stanzas are a hybrid
of John Holland and a blackbird.

XI
For the nectar of John Henry
is an inner richness
like the comet
orchid's and like the sphinx
moth he sports
transparent wings.

XII
A tree becomes part of a sloth's mossy fur.
John Holland's ideas became part of me.

XIII
It was sunrise for 85 years.
It was shining
and it was going to shine.
John Holland stirred
in his bright emergence.

16   Helena Hong Gao

For John's 85th birthday

On 24 Jun, 2013 Michael Arbib responded to Jan's message sent on behalf of Jing and me to the people we invited to make a contribution to this book. His response is the following:

On 24 Jun, 2013, at 2:58 AM, Michael Arbib wrote:

I'd really like a sentence or two from each of you to learn what role John played in your life that led you to the idea for this tribute.

Meanwhile, here's a pass on my entry.

Best wishes
Michael

Michael's request was simple but it made me think a lot about people, relationships of people, and people's roles in other people's lives. What role has John played in my life? Comparing with all the other contributors to this book, I am probably the only one who hasn't known John for long but John has definitely influenced my life, my research interests, and most of all, my understanding of human relationships.

I first met John through Bill Wang in 2005. It was at the workshop called *Language and Genes* organized by Bill in Taiwan Academia Sinica. I went to the workshop a bit late. John was giving a keynote speech in an auditorium and I was sitting in the last row, where I could barely see him, but his voice was so much filled in the air that attracted me easily to listen to him attentively. His agent-based modeling and theory of complex adaptive systems (cas) were new to me. I was trying to follow him by picturing in mind how the model would look like. His examples drawn from economics, global trade, internet, markets, sustainable human growth, ecosystems, immune system, and language all sound like a building plan of a virtual world as in the game of Second Life. How his agents interacted with each other in his cas became vivid and more and more interesting (Attached below are some of John's slides that cover the topics that mostly interested me in the talks he gave in Taiwan, Santa Fe, and Singapore between 2005 and 2006).

Problems Involving Complex Adaptive Systems

Some difficult problems:
   Encouraging innovation in dynamic economies.
   Controlling the Internet (e.g. controlling viruses and spam).
   Predicting changes in global trade.
   Understanding markets.
   Providing for sustainable human growth.
   Preserving ecosystems.
   Strengthening the immune system.

Each problem involves many interacting agents (components) that learn or adapt.

A Complex Adaptive System [*cas*]

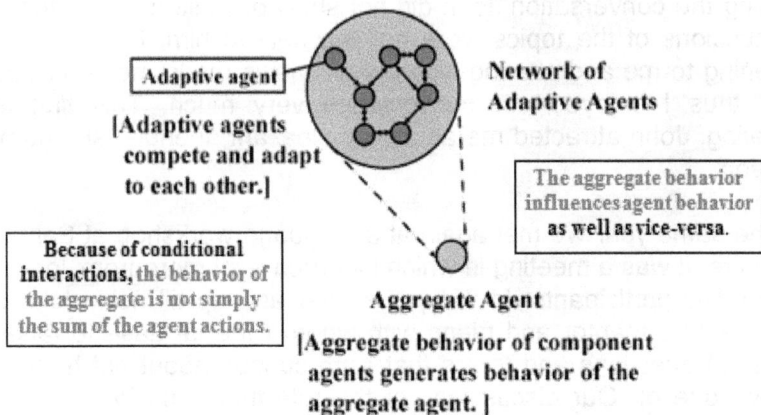

Adaptive agent

[Adaptive agents compete and adapt to each other.]

Network of Adaptive Agents

The aggregate behavior influences agent behavior as well as vice-versa.

Because of conditional interactions, the behavior of the aggregate is not simply the sum of the agent actions.

Aggregate Agent

[Aggregate behavior of component agents generates behavior of the aggregate agent.]

A *cas* Language Model

The major objective is to observe the emergence of a structured language in a situated, multi-agent ('social') model.

The agents have no 'wired-in' (pre-determined) language structure.

The agents are situated in an environment exhibiting "perpetual novelty".

The environment is 'Darwinian' so that agent survival depends upon collecting resources provided by the environment.

The following should be emergent properties:

1) Networks of interaction between agents.

2) Proto-grammatical (sequenced utterance) constructions.

3) 'Meanings' (treated as equivalence classes over environmental states).

4) Increasing complexity and diversity of agents.

John's talk left me many wonders and curiosities. At the discussion session followed by a working lunch, John and I happened to sit side by side. I did not recognize that he was the speaker on the stage in the morning (I could not see his face clearly then from a far distance). We introduced each other and talked freely on several research topics including machine translation and child language acquisition. During the conversation John did not show or make me feel that our discussions of the topics were nothing new to him. His sincerity in listening to me and sharing with me his views made me feel at ease and thus I enjoyed our conversation very much. This first time meeting, John attracted me as a very pleasant, friendly, and humble person.

In the same year we met again at a language workshop at Santa Fe Institute. It was a meeting in which I learned a lot more from John, Bill, and other participants. In the years after and up till now, John has been a true mentor and friend with whom I feel at ease to raise all kinds of questions and topics that I am curious about but have little knowledge of. Our discussions of the role that consciousness may play in child language acquisition have been turned into two joint papers and one research project.

I guess most people including John may have never intentionally planned to play a role in other people's lives, but people do influence each other all the time. I am fortunate to have met wonderful people like John. As friends, we care for each other and share with each other the ups and downs in our lives. We treasure the good time together, the friendship, and the beautiful memories. In my response to Michael's request, I wrote that "the role John has played in my life is mentorship and friendship. His friendliness and humbleness have encouraged me to ask him questions of all kinds and his responses are always inspiring. He has raised my understanding of the scientific topics we have discussed to a higher level".

What John has deeply impressed me ever since I met him are his scientific mind and wide interests. He has widened my mind for science and influenced me on my understanding of what human relationships can mean to an individual, mankind, and the universe. In casual conversations, we sometimes talk seriously about life and death, theism and atheism, but most of the time we enjoy spontaneously teasing each other. These are the most enjoyable part of a relaxed and reliable friendship one can have with a true scientist. Happy 85th Birthday John!

Below is my introduction of John for his speech at the conference *A Crude Look at the Whole* at Nanyang Technological University in March 2013.

Good Morning Everyone!

I am very much honoured to chair this session for Professor John Holland's talk on his newly published book **Signals and Boundaries**. Let me give you a brief introduction of Professor John Holland first.

Professor John Holland studied physics at the Massachusetts Institute of Technology and received his bachelor degree in 1950. He then studied mathematics at the University of Michigan and received his Master Degree in 1954. In 1959 he completed his Ph.D. program at the University of Michigan and became the first person to receive a Ph.D. degree in 'Communication Sciences'; MIT and UM were the only universities that had such programs and in both cases they later became departments in computer science. He is now a professor of psychology and a professor of electrical engineering and computer science at the University of Michigan.

Prof. Holland is a member of the Board of Trustees and Science Board of the Santa Fe Institute. He received the MacArthur Fellowship in 1992, and is a fellow of the World Economic Forum.

In 1975 he wrote the ground-breaking book on genetic algorithms, *Adaptation in Natural and Artificial Systems*.

In 1995 his book *Hidden Order: How Adaptation Builds Complexity* (1995) was translated into Chinese and became a popular scientific book in China.

In 1998 he published his book *Emergence: From Chaos to Order*.

Last year, 2012, he published his new book *Signals and Boundaries: Building Blocks for Complex Adaptive Systems*.

While preparing for my introduction for this session yesterday, I did a Google scholar search of the citation of his ground-breaking book. It showed 17 versions or reprints. By checking the 1992 version, the citation shown was 37398 times.

All these may have been known to most of the audience today.

People in the audience, especially our NTU students may wonder how a professor could become such a great scientist. Whenever people ask John a similar question, he always says that it is because he was fortunate to have good professors. They gave him opportunities to do what he liked when he was a student. The flexible system at MIT at that time allowed him, a student in physics, to do his B.S thesis with Zdenek Kopal, an astronomer teaching 'numerical analysis' in Electrical Engineering, who gave him access to Whirlwind, the first 'real-time' computer being built then at MIT. Because of that experience, IBM recruited John for the planning group for the first commercial computer, the first electronic programmable commercial computer. We can see that John's curiosities and gained experience set the interdisciplinary part of his research at a very young age.

What is also indeed true is that John has a mind that is interested in almost every subject in nature that is complex. His solid theoretical work seems to be all based on natural sciences such as logic, mathematics, physics, biology, ecology, etc., but what made him able to see the relations of the complex parts in a system in the real world was due to his profound knowledge and broad interests of all kinds. For example, his interest in linguistics and communication studies started in the early years of his career and he has never given it up. Over the years he has organized several language workshops together with William Wang at Santa Fe and Asia that have influenced him and a number of us in our research directions in linguistics.

(To me, research in language acquisition is like looking for a lost key to the mystery of a human's ability and a disability or handicapped ability (in the sense that we are never able to learn a foreign language as well as a native speaker when we reach a certain age.) To find this lost key, we need to see the fact that language by itself is a complex adaptive system on its own and our exploration needs to take a multi-disciplinary approach to look at old issues from a new perspective.)

As a person, John is talented in many things. He loves music and literature. He took flying lessons in high school (because he was inspired by his mother, who learnt to fly when she was 45). He tried Chinese calligraphy, practised Chinese painting, plays GO and other board games. He also plays poker with a group of professors at Ann Arbor regularly. Out of his interests, many of his interesting ideas emerged for solving his research questions. Even his sense of humor together with his sketching and painting skills allowed him to be a cartoonist for the MIT Humur Magazine during his years as an undergraduate student. We now understand why he prefers to make all the drawings of the figures and diagrams in his books all by himself.

Another thing that the students in the audience may like to know is that John is also a great professor and a good listener. He would never say "Is there anything else?" to mean "You have asked too many questions. You should go now.", which was mentioned by an NTU visiting professor at his recent talk that some professors do so when they are bored with students' questions.

Chairing John's talk on Signals and Boundaries

With Prof. Holland, you can ask whatever and however many questions you have. You can even trust him to give you credit for your crude thoughts, no matter how immature you think they are. He can always see some reasonable connections and understand why you think so and what can be generated from them. Over the years working with John, I have exchanged with him quite some of my immature ideas, like sound symbolism, consciousness in newborns for language development, and so on, but each time John took it seriously and raised further questions.

As a conclusion, I may say that John's mind is actually a great model of a complex adaptive system. It is constantly upgraded, constantly adapting to the changes of the world, and constantly generating new ideas. Now here is a good example: Soon after his new book *Signals and Boundaries* was published last year, he started writing another book, of which he has completed the first draft and that will soon be published by Oxford University Press.

Before he is too quick for us to catch up with him and move on to writing another book, let's welcome John to give his talk on his current new book, *Signals and Boundaries*.

17   Dave Goldberg

ThreeJoy Associates, Inc.
deg@threejoy.com

Meeting John Holland & 3 Lessons Learned

John Holland has had an enormous impact on my life, and I'd simply like to tell the story of how I met him and a few things I think I learned that had roots in that first encounter.

Meeting John

It was the fall of 1980, and I had returned to Michigan to get a Ph.D. in Civil Engineering (in hydraulics & hydrology). Prior to returning to school, I had been working at a software company in Carlisle, PA that was installing real-time pipeline simulation software in oil and gas pipeline systems, and I had the realization that long-distance gas pipeline operators "drove" pipelines like you or I drive a car. At the same time, I was also reading Doug Hofstadter's book, *Gödel, Escher & Bach,* so like many others at the time, I returned to school with dreamy visions of doing artificial intelligence. When I got to Michigan, I signed up for the standard computer science course on *Artificial Intelligence.* I went to class the first day where I found a sign on the door saying that the course was cancelled.

In seeing the sign, all I could think of was that my dreams of doing AI and pipelines were being crushed by the cruel fates of class scheduling. I did what any eager grad student would do. I leafed through the course catalog looking for something — anything — that might allow me to make some progress toward my idea of doing AI & pipelines for my Ph.D. dissertation. After a good long while, I came across a course called *Introduction to Adaptive Systems, CCS 524,* taught by someone named J. Holland. The course description wasn't exactly what I was looking for — it was filled with a mishmash of weird topics — but it was the best game in town, so I signed up and went to class.

I arrived in the classroom, and standing at the front was an energetic and youngish looking prof. He started by talking about the mechanics of the course. The class didn't have any exams, it had one assignment, a term paper/project on any subject, and it had two course books. One text was something called *Adaptation in Natural and Artificial Systems,* a text written by the instructor himself in 1975. The second book was a collection of papers published in 1963, *Computers & Thought,* edited by Feigenbaum & Feldman. I couldn't help wondering at the time whether there hadn't been anything more interesting published in AI than a 17-year old collection of papers, but I decided to reserve further judgment about the course texts until I could get them both and leaf through them back at my apartment.

After going through the preliminaries, the prof put on his glasses and started talking about what sounded to me like a randomly selected and unrelated series of subjects: genetics, economics, automata theory, schemata, Samuel's checker player, some strange construction he called a classifier system. I'm working hard trying to keep track and make sense of all this, and deep in my heart I wanted him to talk about "real AI." But he seemed earnest enough; there was a deep confidence about him and what he was saying even though, frankly, I didn't understand how any of it fit together, or more importantly, how any of it might lead to a dissertation in Civil Engineering.

That first day in class was quite perplexing, and when I went home, I thought that it couldn't get any worse, but I was wrong. I went to the bookstore got the two texts, and took them home. The collection of papers was dreadfully old and outdated. None of it had working code or anything that would be useful to an engineer trying to apply it. The book even smelled bad — musty and old — like a flooded basement after the water recedes.

And the author's book. Yikes!! It was filled with (a) text passages I had trouble interpreting in any meaningful way, (b) equations that didn't relate to anything I had experienced in my equation-filled engineering education, and (c) no working code, algorithms, or anything concrete that actually worked in a way that a proper engineer might actually apply to something real.

What had I gotten myself into?

But, for reasons I'll explain below, I stuck with the course, and about halfway through I started to understand some of what he was saying. Moreover, I started to dig around a little bit and see if anyone had applied genetic algorithms and classifier systems in pipelines. Couldn't find a thing. I talked to the instructor about the possibility of applying his work in gas pipelines. He was optimistic, confident that the underlying work was sound, but I wouldn't call his response encouraging. To be honest, I wasn't sure if he was humoring me or whether he actually believed that the project I was considering was actually possible.

I approached my advisor in Civil Engineering, Ben Wylie, and he supported the idea of my doing a dissertation connecting genetic algorithms, classifier systems, and pipelines, and in 1981 John and he agreed to co-advise the work.

3 Lessons I Learned from the First Encounter with John

I love telling this story because of the serendipity involved and the tenuousness of the first encounter. As I think back on this story, one of the puzzlers is why I stuck with the course. In many ways, it met none of my prior expectations, and yet I stayed. Of course, I am glad I did. Doing so changed the course of my life, but what was it that kept me in the class and not looking for something else?

I think I stayed, in part, because those first experiences had roots of some of the bigger lessons of what I have learned from John.

Holland Lesson #1: Tell Great, Coherent Stories

I think one reason I stayed with the course is because John told really great stories about his science and his systems. Sometimes the equations were hard to follow, but I always got the story, and it was the coherence of the story that made me believe that ultimately this stuff could be made to work along the lines that he was suggesting.

It's true that a lot of John's work leads to mathematical or computational models of various kinds, but his work always begins with the iterative telling of a great story. My work as an engineer, scientist, and now as a leadership coach has led me to think of stories as the central way in which human beings approach complexity. Stories contain time and spatial relationships, causality relationships, intentionality relationships, clues regarding ontological modes & epistemic certainties and uncertainties all in a compact representation with the possibility of pointers to other stories as well as visual and mathematical representations. I couldn't have had a better introduction to the importance of great narrative in a scientific context than listening to John and his stories about genetic algorithms and complex systems.

Holland Lesson #2: Have the Courage to Jump Long

One of the puzzles for me in meeting John was getting used to his bringing insight from so many different fields. He did it so effortlessly and naturally, yet for a young engineer it was bewildering to see someone jump outside of his own discipline with such confidence and bring back so many useful things.

In the 33 years since that first day in class, university life has become marginally more interdisciplinary, but students of John smile knowingly at each other when others talk about interdisciplinary work. John taught a kind of extreme courage to go wherever you need to go in the interest of doing good work. Rarely do you see it done with such grace and aplomb as by John, but many of us touched by his example are better able to jump longer and further afield than we otherwise would have without his leadership and example.

Holland Lesson #3: Trust Yourself & Those Who Work with You

I think another thing that kept me in class during those first days and weeks when I was still full of doubt and skepticism was John's demeanor. He was completely at ease and quietly confident about the things he said. It wasn't arrogance. It was what coaches might call "leadership presence." He was connected with us in class. You trusted what he was saying even if you didn't fully understand it, and he was curious about our questions and reservations, OK that he might learn something from our challenges and inquiry. It was attractive in a way that was and is hard to describe.

This attitude carried over to the way he "managed" his Ph.D. students. Basically, he trusted us to figure things out. He would listen carefully to our results, he would ask a few questions, but he rarely was directive about what should be done next. At the time, I think I wished for a little more advice. In hindsight, I am grateful he did things as he did.

That's the Way You Do It

I hope this short piece conveys some of the many ways I feel grateful for having met and been influenced by John Holland.

I last saw John in early 2013 in Singapore at an event at Nanyang Technological University. As I sat in the audience, I listened to John discuss portions of his 2012 book, *Signals & Boundaries*. The vibrancy of his storytelling, the courage of his long jumping, and the connection of his presence with & trust of the audience came through as always, and his talk sent me off into a weeklong reflection on *lever points,* reflections reminiscent of ones much earlier. He didn't seem that much different than the youngish looking prof I had encountered for the first time 33 years earlier. And maybe in that final observation is another lesson for us all.

18   Ellen Goldberg

Celebrating John

I was first introduced to John Holland when I was appointed President of the Santa Fe Institute (SFI), January 1996. I had previously learned about John's research accomplishments through fellow University of New Mexico faculty member (Stephanie Forrest), among others, who graciously spent time with me over coffee and lunch explaining genetic algorithms. So before even meeting John, I was impressed by his extraordinary insight and accomplishments regarding computational thinking. As an immunogeneticist working in the field of cellular immunology in a research laboratory throughout my career, John's way of thinking was new and extremely exciting to me. In fact, I wasn't certain, before then, whether or not I would accept the Presidency of the Santa Fe Institute, since so much of the work being accomplished there was based on interdisciplinary theoretical and computational research that was unfamiliar to me. Because of my research background as an experimental immunogeneticist, I wasn't certain SFI was the right fit for me or whether I was the right person to lead the Institute into the next millennium. However, after learning about and studying the research that John initiated, I intuitively knew that the opportunity of being associated with John and others at SFI was a once in a lifetime experience. So, in an indirect way, John helped me make the right decision of accepting this incredible position.

Well, then I met John at an SFI event, even before I started in my new position. Because of all his accomplishments and accolades, I didn't expect to find such an open, friendly and wonderful person like John! (I actually expected to find a somewhat formal and reserved scientist). John immediately won me over with his first smile. His enthusiasm for research and his open and thoughtful ways regarding science and life in general was infectious. We all know about his research accomplishments, but what many people who haven't interacted with him might not know is his talent as a teacher and mentor to junior researchers. He is a gracious and understanding teacher and, in my mind, there is nothing more important. His insight into interdisciplinary research that he imparts to others is a gift that few scientists possess.

John participated in numerous workshops at the Institute and also organized workshops at the University of Michigan to which he invited a number of us from the Santa Fe Institute. He has exceptional taste in identifying the brightest scientists worldwide. And he would include not only researchers who had already made a name for themselves, but, also those who were just at the beginning of their careers. This exposure for young researchers and the time he spent with them was a learning experience in interdisciplinary and computational research that changed the way of thinking by many of these bright young students (and some of us who were more senior researchers but who hadn't been exposed to this way of thinking). Many of these students are now senior researchers who have made names for themselves within their research areas and who are working with the next generation of junior scientists.

One student who was influenced by John and continues to interact with him is Han Jing. I remember her excitement at first meeting him and she recently told me how she continues to interact with him and how much he influenced her life and career. It is a wonderful story. Many of his students have told me similar stories over the years.

Throughout my career I have met and interacted with extraordinary researchers and teachers. John embodies the best of the best — he makes time to teach, learn and interact with all of us who approach him with questions. He also provides thoughtful and insightful guidance for which I will always be grateful.

Thank you, John, and Happy Birthday!!

19 Jing Han

hanjing@amss.ac.cn

A Friend walked out from the book

One of my most lucky and exciting things in my life so far, is to know, meet and become a close friend of John.

Know John from a Book

It was spring 1998, I read the book *Complexity* written by Waldrop. This book changed the path of my life. At that time I was a graduated student major in computer science in China and was quite confused about the future direction of Artificial Intelligence. I was thinking about how to make a program to be smarter than the programmer and that can output something which will surprise the programmer. The idea of emergence and complex systems, especially John's framework of Genetic Algorithms and Classifier Systems, shed lights on my dark world of mind. John provided such an elegant and clever framework that shows strong learning and adapting ability, I think it is the closest work to open-ended evolution. Then I decided what to do in the future: complex adaptive systems. The Santa Fe Institute (SFI) became the top 1 of my dream places in the world.

First Meeting with John

I was so lucky that my dream came true so soon: two years later after I read that book, I applied and got the international fellowship from the Santa Fe Institute (2000-2002). As a Ph.D. student, I visited SFI in January and February of 2001. There, I had a lot of impressive and unusual memories: the first time I

My first meeting with John at Santa Fe. 2/2001

see the western country, the first time to stay and eat western food in an American lady's home, the first time to live in a single lovely adobe house....and the first time I saw SFI, my dream place! People there were so nice and friendly to me. Ellen Goldberg, president of SFI at that time, took me to Washington DC for a trip which was very impressive. I met many people from the book. And one of the unbelievable things was: I was arranged to share the same office with an amazing person: George Cowan — the first president and one of the founders of SFI! I was so thrilled and excited. But the most wonderful part was my last two days in SFI during that trip: meeting John Holland!

Thirteen years passed, but I still clearly remember the scenario of the first meeting: He was sitting in front of a computer at the corridor to check his email. I saw him — an old man with red face, who very much looked like the photo I had seen. At the same time when I was looking at him, he stood up and also looked at me, smiling, and he did something I would never forget: he bowed to me. First I was shocked and then amused. All of a sudden, I felt very close and relaxed with this great scientist — the hero of my scientific world. Then we had a long conversation about research and many other things: Chinese culture, Chinese mountain (John likes mountain Huang very much), Chinese paintings, etc. As an unknown Ph.D. student, I felt so well when a great scientist listened carefully to my research and always commented from positive angles. I didn't feel that I was talking to a senior person with authority. What is more, he was so curious about people. He made me think I am an important person. He even asked me to send him a copy of my wedding photos with traditional Chinese dress. He made jokes all the time and laughed loud! It was just like talking with friends. The conversation was very smooth and joyful. To me, the hero came down from the book, talked and listened to me. I felt extremely excited and happy, and was looking forward to the next meeting.

Friendship

The second meeting was a year later, March of 2002, when I visited SFI again. John invited me for dinner. In the parking place, he walked to my side, and opened the door of the car for me, very much a gentleman. That, again, surprised me and made me feel flattered. He said only the old generation of American did that now. We ate at a Japanese restaurant. Sushi is my favorite. But I almost forgot to eat, because I enjoyed the conversation with John so much. We talked about culture, language, movies, Chinese games such as go and chess, Chinese medicine (Qigong, acupuncture and cupping), garden arts, and even my sweater! He is curious about everything so it is very easy to find a topic to share. It was 10pm when we left. He opened the door of the car again, and asked for tips. In return with humor, I gave him several Chinese coins, which amused him. He sent me back to the place I stayed — Laura Ware's house. Laura saw me and said 'Jing, you look very very happy! You must have had a wonderful dinner.'

It is not easy to express my mixed feeling at that time: admiration because he is a great scientist and he has been doing research that targeted my ultimate goal; closeness because of his great personality and he acted like a father and a trusted friend; curiosity because he is John Holland in the book and he is a foreigner with different culture... I was totally fascinated by him. And he also told me that 'I felt that we could have talked on for days and days' and 'It is so surprising to me that it is as if we had known each other for many years.' It is a delight that both of us don't have to make any pretenses when we talk together.

We exchanged emails frequently afterwards continuing our conversations, which accompanied the period of my dissertation writing and defense. That made me much more eager to continue my research on complex systems after getting my Ph.D. in computer science. So I did continue and became a joint-postdoc of SFI and the Complex Systems Research Center at the Academy of Mathematics and Systems Science of Chinese Academy Sciences. He encouraged me to take risks and pursue the questions in my heart. He told me to do interdisciplinary approach, "Be patient, acquire a patron, and train many students". He said "most posts in large universities are offered to people with large numbers of publications. Departments rarely see very deeply when hiring, using only the most obvious measures, often to their detriment. But it is not always that way. In my case, I had an important advocate, Arthur Burks. He stood up for me in the department and made it possible for me to follow my curiosity, rather than enter the battle of publications. If you don't care about becoming famous, but rather want the chance to follow your ideas, that works well. And, incidentally, because fame depends upon original ideas, not on publications, you are probably more likely to become well-known in the long run as well. It would never have bothered me not to be well-known, as long as I had some students that thought my ideas were interesting. So now I tell my students it is important to have a "patron", much like the medieval noblemen that supported Leonardo da Vinci, if you want to follow your ideals."

I remember one of his students gave a talk in his 80th birthday conference and made some funny statements: John's students were spoiled by him, and found difficulty to adapt after graduation.

Except research, he was like a door for me to explore a much bigger new world for everything. He is the first one to teach me driving, kayaking, how to set up fire of a fireplace, walk on snow with snowshoes, etc. I felt safe to ask any question to him, including the culture difference. And he too. For example, before his first visit to China in October of 2002, he sent me an email and asked me whether it would embarrass me if he hugs me on seeing me, or just shake hands with me according the Chinese culture. I found this funny and told my husband, he gave an also funny answer: just wave hand and say hello.

Another story is about language. He likes to learn Chinese, but it is not enough for our conversation. English is not my mother language, and I only began to use it from the first time I visited SFI in 2001. In the beginning I was so stressed and I had to pay a lot of attention when people were talking. Over the first few days, I was even speaking English in all my dreams! I sometimes couldn't understand what people said and felt frustrated. But with John, I feel relaxed. I can ask him to spell the word and then he would wait for me to look it up in the dictionary at hand. I remember once he told me in a funny way that he felt he was taking advantages when talking with me, he felt himself smarter than me while talking in English. That made me feel totally relaxed.

John learns Chinese calligraphy, 2003

Because we were curious about each other, so the conversations between us were mostly 'question-reply' style. I asked questions about his research, his career, his first book writing, his ideas about the other famous scientists, his family, his childhood, and the culture, etc. He was mostly curious about Chinese culture, arts and history. He tried to learn calligraphy from me. He knew many things of a wide rage and has very good memory.

More and more, I think John likes 'small people', not 'big' ones. Power and control are not his pursue. He said "I'm convinced that it is "small things" like this that make for a happy life, particularly if they are the seasoning on the "main course" of research". He enjoyed equal interaction with people, making jokes and talking about details of life. He does not particularly enjoy being a star under the spot light, nor does he like being treated as a king. His mind is always open and he is more curious at the age of seventies than many young people I have met. He has the innocent heart of a child, which we call "赤子之心" in Chinese. There are very few things he would take for granted. He never thinks of himself as a superior. He respects people and treasures the gifts from 'small' people. When he visited China, he met a lot of fans — most are young students. He was very patient and nice to them. They gave him gifts. One gift was a very common thing that you can easily find in a market. But I saw John seriously pack it into his little suitcase and take it back to US. The next time I visited his house I saw the gift sitting on his bookshelf.

Trips between US - China

From 2002, John had his first visit to China, to attend a workshop 'Adaptation and Intervention on Complex Systems' organized by SFI and the Academy of Mathematics and Systems Science of Chinese Academy Sciences in Beijing in Oct. I hosted John after the workshop and took him to Xi'an.

John in Purple Bamboo Park, Beijing, 2002

John in Xi'an, 2002

Then, I spent a half year in SFI in 2003 for my postdoc, John was my coordinator. He sometimes flew to Santa Fe during that period. I still worked on my own research, but had a lot of discussions with John. John never imposed on me any idea on research, not to mention that he would give me the pressure of publication — good and bad — hard for survival without publications. So I had a lot of freedom and fun there.

During that period, I also got the chance to visit Ann Arbor and his lovely house at the edge of Lake Michigan. I saw that John, at the age of 73, sailed a boat in Lake Michigan. He was much more strong and skillful than I had expected. He was so healthy and energetic, even though he was lazy in doing sports and ate very little. I think this is because of his happiness in heart. He is happy so he has good temper. He has good temper so his body is harmonic. Not like many others, outside evaluation is not John's way to prove his value and esteem. So fame is not so important to him. When John took me to a GECCO conference, I suddenly realized that this low key old man

John at Lake Michigan,
winter, 2003

John at Lake Michigan,
summer, 2003

A working group in AMSS,
Beijing, 2006

SFI Summer School in
Beijing, 2005

has an empire. Most of the participants were John's students, or students of his students, and so on. But he treated that lightly. He said the joy of doing research is like the joy of creating and playing new games with his fellows in childhood. From the words of John, I knew a bit about his parents. I think they gave him a happy and safe childhood, as well as the interest of interaction with people, good temper, and courage of taking risk.

I returned to China and soon became a faculty in the Complex Systems Research Center at the Academy of Mathematics and Systems Science (AMSS) of Chinese Academy Sciences. John visited me and my institute again in 2004, 2005, 2006, 2009, 2011 and 2012 (I became mother in 2007 which kept me super busy for two years). Every time he visited my institute, he would give a public lecture and we would have a group discussion with my other colleagues (Prof. Lei Guo, Dr. Zhixin Liu, Dr. Jiang Zhang, etc.), mainly focused on the recombination of adaptive control and complex adaptive systems. Then a project of game-based approach on adaptive control and complex adaptive systems was carried out for years. John is the consultant of the academic committee of our research center and the member of the International Academic Advisory Committee of AMSS. John participated and taught courses in the SFI-China Complex Systems Summer School in from 2004 to 2006. He was the co-director of the summer school in Beijing in 2005, while I was the associate director of that summer school. I also visited him at 2005 (SFI) and 2012 (Ann Arbor). Except that, we met each other in Singapore in 2009 (his 80th birthday conference) and the conferences organized by Jan Vasbinder in 2011, 2012 and 2013. During those meetings, I got to know Helena Gao, Jan Vasbinder, Tian Qing, and a lot other nice people. John is a big hub in my social network. He has greatly enlarged my world.

John playing Majiang with my family, Foshan, 2004

John also visited my hometown Foshan and my parents in 2004. John said Foshan is the coldest place in the world (which is not true, Foshan has the subtropical climate), because he was walking for hours in a rainy winter day without warm trousers, when visiting the dragon kiln. He needed to drink hot milk and put himself inside the warm blanket after we were back home. It was his first time to play a very popular Chinese game: Majiang. He enjoyed it a lot even it took him much efforts to recognize the Chinese characters on each piece. He was very interactive with my parents and my relatives even just by laughing and gestures because they share no common language. We took John to a vegetarian restaurant inside Reshou Temple. They have many dishes which mimic meat, sea food, very tasty. John usually does not like vegetable except mushroom, but he said it is best restaurant in world he had ever tried.

There is so much to tell about John, one of the very few people that had such huge impact on my life. On the other hand, I saw his happiness and sadness during these years. John has become a wonderful part of my life. I know it, I feel it, and I treasure it.

The last photo is my most favorite photo of John and me. It was taken in Lake Michigan, 2003. We made the sand castle together, we are exploring the world with strong curiosity, enjoying it mostly because it is fun, just like children playing games.

20 Douglas Hofstadter

Center for Research on Concepts and Cognition
Indiana University
Bloomington

So Fine

My contribution to this volume consists in a birthday poem written a few years ago, enhanced with a short set of annotations that I have here provided in order to make this piece into a stand-alone contribution. So here we go.

Four Score and Zero Years Ago...

October 2009

• • •

I sing of John, my pal, my matey,
My friend, my chum — but not just mine!
The John-o-sphere is wide and weighty:
It leaps the Huron, spans the Rhine!
John's gang of friends is grand and great. He
Is loved from Rome to Nome to Haiti,
From Santa Fe to Liechtenstein —
In Holland, even, I'd opine!
John's optimism is innate. He
Enjoys life's crazy, swerving line;
For him, what's random is divine!
And so today, as John turns eighty,
Let's raise our cup and hoist our stein
To toast John Holland — he's so fine!

• • • • •

• • •

• •

•

This piece of light verse was written in honor of my old (yet still young) and dear friend John Holland on the occasion of his 80th birthday; more specifically, it was penned as my small contribution to a symposium that was held at the University of Michigan on that august occasion, in early October of 2009. (For the sake of historical accuracy, I must point out that this symposium took place a few months after John's actual birthday, which is in February, but what's a few months between friends?)

For the purposes of this volume, I would like to append to my ditty a few explanatory words (where "a few" in this case means 1052, give or take a few), first focusing on form and then focusing on content.

First off, I suspect that few readers of this poem will realize that it has the precise form of an *Onegin stanza*, which is to say, a 14-line poem in strict iambic tetrameter, of a sort invented and used to the hilt by Russia's most celebrated poet, Alexander Sergeevich Pushkin, in his deeply beloved novel-in-verse *Eugene Onegin*, which has been translated into English by a number of admirers, including myself — and when I say "translated", I mean translated *in verse*, thus respecting and reflecting all of the Russian original's extremely precise metrical and rhyming constraints. For interested readers, I would particularly recommend James Falen's beautiful, lyrical anglicization of the work.

An Onegin stanza always has the rhyme scheme *ABABCCDDEFFEGG*, and it necessarily features so-called *feminine* rhymes on lines 1 and 3 (the two *A* lines), lines 5 and 6 (the two *C* lines), and lines 9 and 12 (the two *E* lines). (A "feminine" rhyme does not involve words denoting females, nor does it have anything to do with stereotypical feminine traits; the term simply means that the rhyming action takes place in the final *two* syllables of the lines involved, with the penultimate syllable being stressed and the final syllable being unstressed. Two simple examples are "nation/station" and "plastered/bastard".) All the other lines of an Onegin stanza necessarily feature so-called *masculine* rhymes. (Once again, such a rhyme has nothing to do with manliness or sexuality; the possibly misleading technical term merely means that the rhyming action takes place on the *final* syllables of the lines involved, and that those syllables are stressed. Two simple examples are "girl/pearl" and "drunk/junk".) A corollary of this defining feature of the Onegin stanza is that the syllable-counts of the 14 lines always exhibit this exact pattern: 98989988988988, since feminine lines always have one "extra" final unstressed syllable. (Masculine lines might well be envious of this extra feminine appendage.)

In my poem to John, I took the liberty (or rather, I imposed the tightly constraining shackle on myself) of making all six of my feminine lines rhyme with one another (the quintessential one of which ended in the word "eighty"), and likewise of making all eight masculine lines rhyme. There are thus only two distinct end-rhymes in the whole poem. Jumping through all these hoops at once was not a trivial feat! But in order to properly laud John in verse, there was no question in my mind: it simply had to be done!

One last rather nitpicky word about form. In English versification there is a strong tradition of trying to avoid so-called *rimes riches* (that is, rhymes involving a repetition of an identical sound, such as "son/sun"). Some astute readers, aware of this convention, might point to my use of lines ending in "Liechtenstein" and "hoist our stein" as a flaw, since both lines end in the exact same syllable "stein". However, to such careful readers I would respectfully point out that appearances (in this case, spellings) can be deceptive, and in this particular case they definitely are so. Specifically, the "s" in "Liechtenstein" is pronounced as "sh", whereas the "s" in "stein" involves an "s" sound. Case dismissed!

And now a few words about a few perhaps slightly obscure allusions in the poem.

Many readers may be aware of the geographical fact that the Huron River passes through Ann Arbor; fewer, though, may realize that John's lovely home is located on Huron River Drive.

John Holland's last name happens, quite fortuitously, to coincide with one of the appellations of a certain diminutive yet extremely important northern European country, and in penning my verse I noticed and took shameless advantage of this little fact.

The reference to the beautiful town of Santa Fe is probably clear to all, but lest some reader not realize it, John was one of the earliest members (circa 1984) of the then unknown but today world-famous Santa Fe Institute for the study of complex systems, and he has long served on its Board of Directors. Indeed, I personally have always thought of John as the heart and soul of the Santa Fe Institute.

My very brief tip of the hat to John's deep respect — even reverence — for life's crazy, swerving, and random nature is an allusion to his very early recognition (perhaps in the mid-1960s?) of the central and indispensable role of chance (aka "stochasticity") in the short-term and long-term course of evolution, and to his enthusiastic embracing of this core principle in his pioneering work on the computational simulation of evolution, known as "genetic algorithms". I am quite sure that this fact is abundantly saluted and celebrated in some of the other contributions to this volume, so I'll leave my comments on the key role of randomness in life at that.

I would like to conclude by commenting on my choice of the word "optimism". I don't think I have ever known anyone with a more optimistic and sunny disposition than my friend John Holland. John has a magical and amazingly infectious smile, and a delicious and contagious way of chuckling, and smiling and laughing are probably his most frequent facial expressions. John is able to see what is funny about any situation at all, and he takes great pleasure in sharing that sense of drollity with others. But John not only is full of mirth; he is also brimming with graciousness towards others. I have seen John Holland run into intellectual "enemies" at conferences and greet them with genuine pleasure and treat them with genuine respect, stressing what his ideas and theirs have in common. He has a supreme knack for seeing value in the ideas even of those people who he is most at odds with. I have always been astonished at this marvelous, enormously generous and flexible trait of John's mind and personality. I think that in some fashion this fact is responsible for the fact that "The John-o-sphere is wide and weighty" and that "John's gang of friends is grand and great". He is one of the most human of people I have ever known, and one of the most modest as well. I am so lucky to have met John Holland and to be able to count him among my close friends. John is, indeed, so fine.

Post scriptum

My wife Baofen, ever the perfectionist, after reading the poem above, asked me why I'd used a poem that was five years old. I said I thought it was cute and appropriate, but she seemed to think I should have written something new for this occasion. And so the next day, thinking it over in the shower, I said to myself, "Okay, what the hell — why not give it a shot?" And so in the sweet solitude of my shower I tried to think up a feminine end-rhyme that could be used six times in a celebratory fashion, and a masculine end-rhyme that could be used eight times. Not obvious! But while still underwater, I came up with "ever" for the former, and "ive" for the latter. And as soon as I emerged from my ritual self-dunking, I went straight to work, giving it the old college try. Here's the result:

Forever Thrive

For February 2, 2014

• • •

Dear John, it seems, soars on forever:
Still going strong at 85!
His earthly bonds God's failed to sever;
Indeed, John's jauntily alive
And kickin', and still bloody clever!
Like Archimedes with his lever,
John moves the earth — in overdrive!
To be so youthful we should strive;
To have such pep we should endeavor;
To foil old age we should connive,
And jump like John, with joy and jive!
Perhaps his secret's "Ne'er say never!"
In any case, this toast now I've:
"Till he's 100, may John thrive!"

• • • • •

• • •

• •

•

21   Erica Jen

"For John Holland"

It was because of John Holland — actually because of his classifier systems — that I became first interested, and later actively involved, in SFI. For me, that puts John and his classifier systems high on the list of favorite extended phenotypes — together with beavers and their dams, bower birds and their nests, spiders and their webs — that enable organisms to alter their environments, to modify the behavior of other organisms, and to have the power sometimes to bring about great change at immense distances. Certainly, as one tiny example, the duo changed my life forever.

Here's how I got caught in John's web: In 1988 I was invited by Dan Stein to give a one-hour seminar on cellular automata at the newly established Summer School for Complex Systems in Santa Fe, NM. The seminar was on a Monday afternoon — but I showed up at the school early to hear John talk on classifier systems. The talk was so good that I returned on Tuesday, having persuaded my husband George to attend as well, and with each subsequent lecture from John, we became more intrigued not only by the classifier systems but also by the thought processes of the individual who developed them.

The clincher for me came when John put forward the possibility of emergence of stable hierarchical rule structures in classifier systems. As I remember, he offered some plausibility arguments, but he didn't push them as most others would have. There was no overreaching, just a deeply thoughtful combination of analysis and intuition as to what might be the essential building blocks of what he gently suggested to us as a possible example of a "major evolutionary transition." John's vision and commitment to the fundamental scientific question — rather than to his own approach or his own constructed solutions — were irresistible. And it wasn't long before I came to realize that if SFI was the kind of institution that valued and encouraged John's way of thinking, then SFI was the place I wanted to be as well.

I have one story that falls into the no doubt otherwise sparse category of "Nightmare Scenarios Featuring John Holland." The context was the very first workshop I helped organize at SFI; my co-organizer was a researcher who was a complete newcomer to the Institute. Much of the early workshop planning was done by email and involved even more than the usual wide-ranging tossing about of relevant and irrelevant suggestions for workshop topics, participants, and goals. Late one night, I received an email from my co-organizer that contained a rudely dismissive reference to the use of genetic algorithms, and approaches like the classifier system, for modeling of biological processes. Together with a colorfully emphatic diss on John Holland as the pioneer in that area.

The email message was completely inconsequential, except that.... I forwarded it by mistake to John. A panicked thirty minutes later, I had managed to track down John's computer systems manager at Michigan and was pleading with him on the phone to delete the message before John read it. To the systems manager's credit, he refused. And anyway, he told me, "John has already read it."

Actually, that's pretty much the end of the story, which is what makes it remarkable. Next day I talked with John who laughed it off, and reaffirmed his interest in attending the workshop so as to learn the detailed biology of the pathways for which his modeling approach was deigned to be so useless. Toward the end of the workshop — at which John had not asked for a speaker's slot since he wasn't sure if he had anything to present — John gave an impromptu blackboard talk proposing a computational modeling approach to some aspects of the biological phenomenon being discussed. It was of course a great talk.

One more thing I want to include in tribute to John. It's a poem called 山中問答 ["Green Mountain"] written by 李白 [Li Bai], a poet from the Tang Dynasty whose work John knows well. I think it fits him.

問余何意棲碧山，
笑而不答心自閒。
桃花流水窅然去，
別有天地非人間。

You ask me why I live on Green Mountain –
I smile in silence and the quiet mind.
Peach petals blow on mountain streams,
Not among people, a world apart.

Erica Jen
October 2013

22   Stuart Kauffman FRSC

The Seminal John Holland

April 16, 2014

It is an honor to write a chapter for my old and admired friend, John Holland. I start, if I may, with my first defeat by John, before I met him. I managed to explore the behavior of Random Boolean Nets with one input per node and even proved some theorems, with young pride, only to learn that John, of course, had preceded my by a number of years. John, how much you have taught so many of us.

A number of us, John, Brian Arthur, Chris Langton, myself and a number of others were extremely fortunate to share the first decade of the Santa Fe Institute, from its days in the adobe nun's quarters with pink bathrooms, to the modest offices to its now wonderful Cowan campus.

A few words about SFI in its early days. SFI was formed in 1984 by George Cowan and other Senior Fellows of Los Alamos National Labs, with early participation by Murray Gell-Mann, Philip Anderson, and Kenneth Arrow, three Nobel laureates to add power, David Pines, and a growing swirling coterie of younger scientists.

In 1986 Jack Cowan and Marc Feldman organized an SFI meeting titled, by Jack, "Complex Adaptive Systems" a phrase that popped from somewhere in Jack's mind. We met for two weeks with perhaps 25 people increasingly entranced by "complex adaptive systems", whatever CAS were. I like to think that a mixture of passion and utter confusion drove the early fun and exploratory zeal of the SFI. We were exposed to more new areas of science per week than at least I knew existed.

John Holland was everywhere, a dominant, powerful, evocative, winning mind and person. John had recently invented Genetic Algorithms, now so famous and familiar to all of us. Several strands led to thinking about the structure of "fitness landscapes", including the work of Eigen and Schuster, both SFI visitors, Schuster's student group, including Peter Stadler, Walter Fontana, and others, Phil

Anderson and spin glasses with their complex potential landscapes and other work, including work by John and his then recent graduate student, Melanie Mitchell beginning to ask in their case, how the structure of a fitness landscape affected the genetic algorithm's capacity to find good solutions to hard optimization problems. In turn this tied into questions about the effectiveness of recombination, central to the genetic algorithm, on smooth or rugged landscapes. As John told us, there has to be information in the structure of the landscape for good solutions to be found. Later work showed that recombination does not work well on highly rugged landscapes.

I would estimate that John's passion for and command of the Genetic Algorithm drove a quarter to a third of the early work at SFI. Included in this were optimal ways to use the now familiar operators John had invented, the density of "don't care" conditions, the rates of mutation and recombination, the details of the fraction of each population at each generation held over to seed in the same or mutant, recombinator or both forms, on the behavior of the algorithm.

Along with simulated annealing, also emerging at that time, I think the genetic algorithm helped drive the increasingly explored domain of hard combinatorial optimization problems, including those that are NP complete. In short, a large coterie of relevant questions, approaches, and beginning answers, had their foundations laid, in no small part by John Holland.

In the meantime, John, with Brian Arthur, Ricard Palmer and others, constructed a stock market model aiming to test if price bubbles could arise, given a fundamental value. Indeed such bubbles could form, bearing on Rational Expectations in economic theory, the follow on to Arrow Debru Competitive General Equilibrium.

In another line of hard and inventive work, John had introduced "Classifier Systems", in which some hard problem was to be solved using linked IF-THEN chains. The central problem faced by John was "credit assignment". If a given chain was used and a good result was obtained in a periodic reward, how was credit to be assigned to which members of the IF-THEN chain. Given a multiplicity of trial chains, John used posting on a board, and attempted so assign credit. I do not know the ultimate outcome, about which John can speak, but my own sense is the credit assignment problem remains an enigma to propositional AI as in If-Then chains. But that too is fine, hard problems not easily solved point out that the problems are hard. More they may drive search for alternative approaches.

Then John became involved in models of ecology and drove work in ways I do not know well enough to comment upon. His books have reached wide audiences, he finally received the MacArthur Fellowship he richly deserved.

The hope of the early SFI was to establish "The Science of Complexity" as a new field. I think SFI succeeded and has now been copied around the world. John's contributions at the University of Michigan and at SFI to that birth are high. He was and is "The Seminal John Holland".

23   Steve Lansing

A First Encounter with John Holland

In 1992 I accepted an invitation to participate in a seminar at the Santa Fe Institute about emergence. This chance event changed my life. I gave a talk about a simulation model of Balinese water temple networks, based on work with my friend James N. Kremer. Our model showed that the temples were organized into networks with functional significance, enabling dozens of villages to achieve a globally optimal system of water control and pest management. After concluding my talk, I asked for questions. Walter Fontana asked whether the water temple networks could have self-organized. The question was intriguing, and I began to listen very closely to the other talks.

Those talks included Chris Langton on "computation at the edge of chaos", Stuart Kauffman on NK Boolean models and the "origins of order", John Miller on evolutionary games, Charles Taylor on adaptation and John Koza on genetic algorithms. Fortunately for me, my talk came early in the program, and Walter's question pointed me towards a possible connection between these grand concepts and my humble water temples. As one talk followed the next, two things began to come into focus for me. The first was that I was in the presence of genius. I recall thinking that this must have been like coming to the Frankfurt School in the thirties, or the École Normale Supérieure in the seventies...the emergence of a new way to see the world. The second realization came when I blurted out a question, "where did these ideas come from? How did all this begin?"

John Koza and Stephanie Forrest laughed and said simply, "John Holland"! And Chris followed up by explaining the connection between "Adaptation in Natural and Artificial Systems" and his own dissertation. Then the discussion moved on. But later I cornered John Miller to ask for details, and learned about the genealogical thread connecting von Neumann to John Holland via Arthur Burks. When I returned home I began to read John Holland's 1975 book, and with the gracious help of Charles (Chuck) Taylor, began to tinker with adaptive models. This involved a visit to his new Connection Machine at UCLA, and a vision of parallel processing on a supercomputer from the future. Meanwhile, as I learned, tropical biologist Tom Ray was bringing John Holland's creations to life in his marvelous Tierra digital ecosystem. And Andy Wuensche was turning attractor basins for random Boolean networks into shimmering images that seemed to come from somewhere in the Hopi universe. It seemed that in the evolutionary potential of computation, John Holland had discovered some kind of portal. And I remembering wondering what von Neumann would have thought of it all. The following year, I came to Santa Fe for Langton's third conference on Artificial Life, and gave a talk showing that yes, as a matter of fact, Balinese water temples will indeed self-organize.

24   Simon Levin

At an 80th birthday party for Kenneth Arrow, perhaps the nicest compliment (among many) that I heard came from a colleague who said that Ken had never stopped being a graduate student. I think that that description applies equally to John, and that helps explain why both these senior figures are always in the forefront of new ideas and new opportunities, like the Santa Fe Institute. John is one of that rare breed of energizer bunnies, never content to trudge well-worn pathways, but always excited about new ideas, new people, and new pathways.

I first met John at Santa Fe, where he and a number of like-minded visionaries had created one of the most exciting venues for idea exchange and development of new science the world has seen. One of the remarkable things about SFI is the role that younger intellects play in creating an atmosphere, and John has helped cultivate that by mentoring and encouraging the next generations, and advocating for them whenever the opportunity arises. There is of course self-interest there, since John clearly thrives on new ideas, new perspectives, and new colleagues; but the arrangement is mutual, because John is able to involve himself deeply in learning about others' scientific adventures, and to offer penetrating and helpful new ways to think about things. I have since had the good fortune to interact with John in Michigan, in Princeton, in Singapore, and the only things that are constant about those interactions is that there is always some new scientific insight, and there is always a sense of joy upwelling from John in talking about science.

Happy belated birthday, John, and many more. Thanks for all you have done to help create complexity science, for explicating the power of genetic algorithms, for elucidating the importance of complex adaptive systems, and for enriching so many of our lives by your infectious enthusiasm that motivates us to go places we would not go otherwise.

25   Melanie Mitchell

In 1984 I moved from Boston to Ann Arbor to start graduate school in Computer Science at the University of Michigan. I knew almost nothing about the department, and indeed very little about computer science itself. After obtaining a bachelor's degree in math and not knowing exactly what to do next, I had stumbled upon Douglas Hofstadter's book, *Gödel, Escher, Bach*, which convinced me to transform myself into an artificial intelligence (AI) researcher. I managed to join up as a volunteer with Hofstadter's group at the MIT AI lab, and naturally followed along when he moved to a faculty position at Michigan.

During my first year, along with the usual courses in algorithms, compilers, and theory, I signed up for the decidedly unusual course taught by John Holland: "Adaptation in Natural and Artificial Systems", based on his now-famous book of the same name. I had no idea what to expect, but the course ended up changing my life in more ways than one.

The class, like the book, covered John's theoretical framework for studying adaptation in whatever guise it appeared, based on his mathematical approach to the "exploitation versus exploration" tradeoff. The mathematics of this book inspired the genetic algorithm, which is now John's most famous creation, but his motivation was as much to understand biology as to create a new method for AI.

I absolutely loved every minute of this course, not just for its amazing, mind-opening (and mind-blowing) content, but also for the opportunity to see John's teaching style. I can describe this style in two words: infectious enthusiasm. Not only was John's obvious love of the ideas infectious, he also really cared what the students were thinking about. Whenever a student asked a question, no matter how wacky or incoherent it might have seemed to other students, John always said, with utter conviction, "That's a great question!" John then proceeded to dig deep enough into what the person what thinking so that he ended up convincing the rest of us that indeed, it actually was a great question. Everyone in the class was inspired both by the topic and by the value John placed on our input. I learned a lot about complex systems, but also a lot about how to be an effective teacher — lessons that I still strive to put into practice.

As I progressed through Michigan's graduate computer science program, John's impacts on my education increased further: I started attending his graduate-student seminar; I got involved in research on genetic algorithms; and John became the co-advisor with Doug Hofstadter for my Ph.D. project on the Copycat project. John is famously supportive of his academic offspring:

> After I received my Ph.D. (or "got my union card", as John described it), my first two positions (a postdoc in the Michigan Society of Fellows, and a faculty position at the Santa Fe Institute) were largely due to John's influence. I am very grateful to John, for teaching me how to think, for teaching me how to teach, and for still inspiring and encouraging me to ask more "great" questions.

John, Doug Hofstadter, and me just after my U. of Michigan Ph.D. defense, 1990

26   Scott E. Page

Shoshin

"A university professor went to visit a famous Zen master. While the master quietly served tea, the professor talked about Zen. The master poured the visitor's cup to the brim, and then kept pouring. The professor watched the overflowing cup until he could no longer restrain himself. "It's overfull! No more will go in!" the professor blurted. "You are like this cup," the master replied, "How can I show you Zen unless you first empty your cup."[15]

I first met John Holland through his book Adaptation in Natural and Artificial Systems in 1990. Even though by that time the book and the many ideas within had gained traction within computer science and psychology, economists were just beginning to take notice, primarily I think, owing to a series of conferences and papers on complexity coming out of the Santa Fe Institute.

These ideas for a new economics included such heretical thoughts as: Perhaps markets weren't in equilibrium, perhaps people did not act rationally but instead were constantly adapting within a complex system that had the capacity for both emergent order and near spontaneous collapse.

John was among the people posing these challenges, and word within the ivory tower was that John's book was the natural place to begin to educate oneself about complex adaptive systems. So, a group of faculty and graduate students decided to get together and read John's book.

The two most common reactions to the book were (1) this is definitely not economics and (2) this is rather interesting. Given that we were all studying economics, the first reaction had a greater influence than the second on most of the group. But for some of us, the second reaction overwhelmed our disciplinary incentives and we became hooked. We wanted more.

---

[15] Source: http://www.hassou.co.uk/index.php/2010/10/22/shoshin-beginners-mind/

More required meeting the people with the ideas. Complexity theory was moving so fast that many of the cutting edge ideas had yet to be written down on paper. They were all in people's heads. The lack of any meaningful World Wide Web at that time meant that working papers and presentations were precious commodities. So precious in fact, that visitors to the Santa Fe Institute would be told to "be cool" about taking working papers. This meant don't grab every single working paper and stuff it in your backpack. Show some discretion.

Thus, learning more meant going to Santa Fe and Ann Arbor, the two hubs of complexity science. At that time, the personalities at the Santa Fe Institute were larger than life. Not only could one see Nobel laureates like Murray Gell-Mann, Kenneth Arrow, and Philip Anderson, but also people brimming with excitement about complexity science and how it would change the world. A brief conversation with Stuart Kauffman, Chris Langton, or Brian Arthur could not only convince you of the truth of complexity. It might well turn you into a proselytizer yourself.

A trip to Ann Arbor was markedly different. Though equally impressive in their academic and intellectual achievements, John Holland, Michael Cohen, Bob Axelrod, and Carl Simon were much less interested in self-promotion. John in particular stood out for his gentleness and openness to ideas.

In my first meeting with John, I was talking with him about the schema theorem, one of many important results in his book. John began to probe me with questions, to ask me how I thought that genetic algorithms worked. Then he told me to walk him through some simple examples so that he could understand.

He wasn't quizzing me to see if I understood genetic algorithms. Clearly, he knew how they worked. He was eager to learn my way of thinking about how they worked. My particular interpretation was more Bayesian than John's. I suggested that perhaps genetic algorithms adapted their level of variation as a function of the ruggedness of the landscape. He seemed more interested in my work than I was.

John's reaction was unlike that of any great scientist I'd ever met. He seemed more like a really supportive seventh grade teacher than someone who had taken the academic world by storm. A few years later, my friend John Miller introduced me to the concept of *Shoshin – The Beginner's Mind*. A beginner's mind requires openness and an ability to rid oneself of preconceptions. A beginner's mind is eager to learn, and not at all interested in impressing you with its brilliance. To the contrary, even in advanced study, it comes at a problem as though a novice.

John Holland has a beginner's mind.

In subsequent years, I have had the great pleasure or watching John work through some of his ideas, most notably his thinking on niches. In discussions with students and colleagues and in making presentations, John seems to have little interest in making some enormous intellectual contribution, in solving some great open problem, or in contributing to a specific literature. Instead, he's focused on the ideas, on how things work. He's frequently saying thinks like "now that's interesting."

Understanding how things work often requires building and thinking through simple models. John refers to this as inductive thinking. For example, John can spend hours talking about what he likes to call *billiard ball models*. In his conception, a billiard ball model consists of entities that bump into one another and, very much unlike real billiard balls, change as a result of those collisions. One billiard ball might be a chemical reagent. When that ball bumps into another kind of ball, a chemical reaction occurs transforming both balls.

Simple right? And yet, amazingly complex if one thinks through all of the possibilities. The genius of John rests partly in his ability to construct these simple models — billiard balls and building blocks — that prove capable of producing complexity

When the BACH group (named after Burks, Axelrod, Cohen and Holland) was up and running, John would often make presentations where he would demonstrate his style of inductive thinking. He would describe simple computer programs that he had written and he would marvel at the results they produced. He would soon have us marveling as well.

When you watch him think through these models, you at first find them entirely disconnected from reality, as an intellectual playground of purely aesthetic concern. And, then, as if by magic, John can connect his models to chemistry, to brain science, to ecologies, and economies. And you suddenly realize that John (*genus scientist abstractus*) is doing the most relevant work of anyone you know.

As Michael Cohen once said to me, "I've learned long ago to never question the relevance of what John is doing, but to always question what it is that he's doing." It's through this questioning that John learns. On those rare occasions when someone finds an error in John's thinking, John will happily say "that's right." He'll be genuinely pleased. There won't be a moment's anger or frustration at time wasted. There will be no ego that must be assuaged, just a joy in that he's now further ahead than he was.

John may be turning eighty-five, but he's among the youngest people I know. The rest of us, by comparison, are cluttered. Our minds occupy themselves with reputational epiphenomena such as publications and citations, while John continues to play with and advance his brilliant ideas.

More than any of my colleagues, John voluntarily and regularly engages with undergraduates. He teaches an undergraduate class in complex systems and has for years. The students think of him, not as some great scientist but as John who loves to play with ideas. Only after they learn of his academic fame and reputation do they begin to refer to him as "Professor Holland."

At Michigan, during the summer students get together to read books and talk about complexity. More often than not, some young undergraduate will be bursting with enthusiasm and explain to all those gathered exactly how complex systems work and why they're so important. There among people literally one-fourth his age, will be John.

His beginner's mind eager and alert.

27   Mercedes Pascual

One of the great things of moving to the University of Michigan was the opportunity to get to know John and interact with him through the Center for the Study of Complex Systems (CSCS). Perhaps I had this special chance to get to meet John because as I learnt later, he loves the writing of Argentinian writer Jorge Luis Borges, including a very short story presumably about cartography, but really about the challenge of complexity. I once cited this short text as the opening of an essay and application for a fellowship that greatly influenced my career. John was involved in that decision and I owe him enormously for this, even if probably some of my 'luck' had to do with a shared love of Borges. So, in his honor I'd like to include here a translation of this text from Spanish to English:

"In that Empire, the Art of Cartography reached such Perfection that the Map of one Province alone took up the whole of a City, and the Map of the Empire, the whole of a Province. In time, those Unconscionable Maps did not satisfy and the Colleges of Cartographers set up a Map of the Empire which had the size of the Empire itself and coincided with it point by point. Less Addicted to the Study of Cartography, Succeeding Generations understood that this widespread Map was Useless and not without Impiety they abandoned to the Inclemencies of the Sun and of the Winters .... (Suarez Miranda, "Viajes de varones prudentes", Libro XIV, 1658).

— Translation of "Del Rigor en La Ciencia" in "Cuentos Breves y Extraordinarios" by Jorge Luis Borges y Adolfo Bioy Casares.

Thank you John for your many inspiring contributions to charting the study of complex systems!

To those of us who admire the many roads you have opened, you are our intellectual "hero". Especially because in the thirteen years I have spent at UM, since I first met you and had the chance to participate in the re-instated BACH meetings at CSCS (with Bob Axelrod, Michael Cohen, Rick Riolo, Carl Simon, and added, Scott Page and Mark Newman), I have been amazed at the sense of excitement and wonder you continue to bring to the pursuit of complexity science. When I think of where I'd like to be in my work as time passes, it is in this mental state that you so greatly exemplify.

I look forward to continue discussing the new developments of your thoughts on the emergence of niches and diversity in complex systems.

28 Jerry Sabloff

Congratulations, John on your 85th birthday. You have been a key contributor to the Santa Fe Institute's scientific mission since its beginning nearly three decades ago, and your book *Emergence: From Chaos to Order* has a valued place on my bookshelf.

Very best wishes, Jerry Sabloff (on behalf of all your friends at SFI).

John Holland, Geoffrey West, Nina Fedoroff at Santa Fe, NM, 2010

2005 Complex Systems Summer School

July 11 - Aug. 5, Beijing, China

Public Lecture

Title: Genetic algorithms and hidden order

Speaker: **Professor John Holland**

University of Michigan and SFI, USA

Abstract

Complex adaptive systems (cas) consist of many components (agents) that interact in conditional (nonlinear) ways and adapt (learn) as they interact. Many of our most difficult problems center on cas: markets, biological cells, and ecosystems are familiar examples. Innovation and diversity are common, important features of these systems. Though networks are a natural representation of the interactions in cas, the nonlinearities are severe enough that most theorems of traditional mathematics are of limited help.

Despite the mathematical difficulties, there are regularities and a hidden order in cas that can be revealed by careful study. To acquire this insight, we must concentrate on the "building blocks" (standard components) from which the agents are constructed. It is a commonplace that we understand the world around us - be it proteins, spacecraft, or languages - by discovering the relevant building blocks. It is easily established that most innovation comes from combining well-known building blocks in new ways. To understand cas, then, we must understand the ways in which adaptation (learning) recombines building blocks.

Genetic Algorithms (GA's) produce adaptations through the simultaneous discovery and recombination of large numbers of building blocks. This lecture will outline the background and mathematics underpinning the construction of mathematical and computer-based models of cas, concentrating on GA's and associated analytic techniques.

主办单位: 美国 The Santa Fe Institute (SFI)　Co-directors: John Holland 教授
中国科学院理论物理研究所　　　　　　　　　University of Michigan and SFI
中国科学院数学与系统科学研究院　　　　　　陈晓松 研究员
中国科学院研究生院　　　　　　　　　　　　中国科学院理论物理研究所

报告时间: 7月27日 (星期三) 19:30 – 21:00

报告地点: 中国科学院基础科学园区 (五所) 报告厅 (北四环保福寺桥西南角)

感谢国家自然科学基金委员会的资助

108

Terry Jones, Al Gore, John Holland at Santa Fe Institute, possibly in 1996

John Holland in New Mexico (date unknown)

(The above photos and the poster are from Ginger Richardson and the SFI archives, sent in by Ginny Greninger on 22 August, 2013.)

29   Carl P. Simon

Professor of Mathematics, Economics, Complex Systems
and Public Policy
The University of Michigan

A BIOGRAPHY OF BACH AND PROGENY

John Holland begins each of his books with expressions of delight
and gratitude for his regular interactions with the BACH brainstorming
group at The University of Michigan (UM). The BACH group played a
central role in the development of the study of complex adaptive
systems, even before there was a Santa Fe Institute (SFI). When
the newly formed SFI tried to hire John Holland away from the
UM and the BACH group in the mid 1990s, the University began to
understand how special the BACH group was and offered support
that eventually led to the establishment of the University of Michigan
Center for the Study of Complex Systems (CSCS). This paper
describes the history of the BACH group and how it morphed into the
UM Center for the Study of Complex Systems.

I.   BACH

Cast of Characters:

<u>Act One (The First Four, 1982):</u>

Arthur **B**urks (Philosophy, EECS; worked with John von Neumann in
development of first computer, John Holland's thesis advisor)

Robert **A**xelrod (Political Science, Public Policy; pioneering work on
repeated prisoner's dilemma and emergence of cooperation)

Michael **C**ohen (political science, public policy; adaptation in
organizational decision making)

John **H**olland (EECS, Psychology; genetic algorithms, agent-based
modeling (ABM), computer learning)

Act Two (1982-83):

William Hamilton (evolutionary biology; evolution of sex, kin selection, altruism)

Rick Riolo (CSCS; student of Holland, ABM, genetic algorithms)

Carl Simon (UM math, economics, public policy; role of primary infection in spread of HIV)

Act Three (1984-2000):

Reiko Tanese (Holland Ph.D. student; genetic algorithms)

Douglas Hofstadter (Psychology, Walgreen Chair; Godel-Escher-Bach, cognitive science)

Melanie Mitchell (Hofstader Ph.D. student; computer learning by analogy)

Michael Savageau (microbiology; bioinformatics, gene regulation)

Act Four (2000-2014):

Scott Page (political science, complex systems; diversity in institutions)

Mark Newman (physics, CSCS; theory of networks)

Mercedes Pascual (ecology; population and community dynamics, disease spread)

History of BACH

It all started around 1980 with Michael Cohen initiating a conversation with John Holland at the end of one of Holland's classes. Actually, John points to a bit of pre-BACH history:

> The true precursor to CSCS was the Communication Sciences (CS) program (in LS&A) started by Arthur Burks (Philosophy) and Gordon Peterson (Linguistics) in 1956. It was a half-century ahead of its time: It covered all aspects of communication from language and information theory through computer algorithms and programming to neural networks, signaling, and evolution in biology. Each course was taught by a member of the appropriate department, for example: Gunnar Hok (EE-information theory), Henry Swain (Pharmacology), Bernard Galler (mathematics), and so on. As CS moved toward departmental status Bill Wang and I designed the first CS introductory course.

Around 1980 Michael Cohen was auditing just such a John Holland class and realized that they shared a common interest in complex adaptive systems, Cohen from the social science and organization studies side and Holland from the computation side. Cohen introduced Holland to his friend and collaborator Robert Axelrod; eventually (probably in the Fall of 1982, just after Axelrod returned from his sabbatical in Palo Alto) they agreed to meet to discuss common interests. Axelrod had already written papers on the role of schemata in cognition, while a cornerstone of Holland's founding of genetic algorithms was the Schema Theorem. They were soon joined by Holland's mentor, Arthur Burks. The four were immediately excited by this opportunity for cross-disciplinary brain-storming and agreed to meet every other week for two hours. They used their initials and called themselves the BACH group.

Recently, John Holland reminisced on those first meetings:

> When Michael Cohen introduced Bob and me, we met at my office in (then) East Engineering because I had a large office with a conference table. Bob and I found we had many questions and ideas that interacted naturally, and after the first meeting or two, we agreed that Art Burks would be an ideal addition. Thus, BACH was formed, and we met regularly once or twice a month for the next 30 years. It was an informal group and we never looked for direct funding.

Axelrod soon invited UM evolutionary biologist William Hamilton to their meetings — and soon thereafter Holland's Ph.D. student Rick Riolo. In 1981 Axelrod and Hamilton had published an award winning paper in *Science* on the evolution of cooperation, applying Darwinian evolutionary theory to the iterated prisoner's dilemma. During the early 1980s, Axelrod would go on to write the book *Evolution of Cooperation*. Riolo became the BACH group's computer algorithm and programming expert, especially as the group grappled with computer learning algorithms such as Classifier Systems.

Major topics of the BACH group during the 1980s were Axelrod's efforts to incorporate a genetic algorithm approach to the evolution of cooperation in the prisoner's dilemma, Axelrod and Hamilton's work focusing on the evolutionary behavior of parasites as the basis for the evolution sex, Cohen's efforts to formalize the adaptive role of routine in organizations, and Holland's refinement of genetic algorithms and classifier systems with concepts such as tags, look-ahead, building blocks.

To add new fields to the BACH discussion, mathematician/economist Carl Simon was invited to the group in 1983. In 1984, Bill Hamilton left to return to England for his appointment as the Royal Society Research Professor in the Department of Zoology at Oxford, though he would return regularly to brainstorm with the BACH group. When Douglas Hofstadter, *Scientific American* columnist and author of the Pulitzer Prize winning book, *Gödel, Escher, Bach*, accepted the Walgreen Chair at UM in 1984, he was immediately invited to join the BACH group. He and his student Melanie Mitchell brought a new interest on cognitive computing via analogies to the workshops. Hofstadter returned to Indiana University in 1988; Mitchell moved to the Santa Fe Institute in 1992. In 1986, UM microbiologist Michael Savageau was invited to the BACH table, in light of his results on a novel method for characterizing nonlinear dynamical systems in biology, especially regarding gene regulation.

Throughout the 1980s and 1990s, the BACH group continued to meet for two hours every other week. The meetings were so intense and insightful that there were very few meeting absences. Often the group's meetings were dedicated to constructive criticism of research projects of BACH members, including critiques of members' book projects. These book projects included Holland's co-authored 1986 book on *Induction*, Holland's *Hidden Order: How Adaptation Builds Complexity* published in 1995, with the initials of the BACH group cleverly referenced in its subtitle, and Holland's *Emergence*, published in 1998. In 1999 Axelrod and Cohen published the first overview of complex systems for the lay reader, *Harnessing Complexity*, which they dedicated to the BACH group for the inspiration that the group had given their research.

Most meetings, however, were brainstorming sessions on some aspect of adaptive systems. In the 1988-1992 meetings, these topics included:

- Examples of adaptive systems, such a economies, game strategies, Darwinian evolution,
- Obstacles to adaptation, such as premature convergence;
- Useful concepts in general adaptation and in adaptive mechanisms, such as path dependence, schemata, tags, bucket-brigades, and look-ahead;
- General processes in adaptation, such as competition, crossover, and mutation;

- The trade-off between exploration for new knowledge and exploitation of existing knowledge;
- The role of internal models in adaptation.

Other questions that kept the group's attention included:

- What problems are *suitable* for treatment by genetic algorithms?
- What kinds of problems are *best* handled by genetic algorithms?
- How should parameters and variables be chosen for such problems?
- How can classifier systems be simplified and standardized?

Throughout its first years, the BACH group asked for no university funding. However, in 1986 UM President Harold Shapiro announced the establishment of the Presidential Initiatives Fund, with support from the W.K. Kellogg Foundation "to encourage researchers from a mix of UM schools and colleges to undertake a new set of joint research ventures." Spearheaded by Michael Cohen, BACH applied to the fund for a project entitled "Parallel Adaptive Systems in the Social and Biological Sciences." In 1987, they learned that their proposal was awarded a three-year grant, which they used for hardware, summer support and salary for Rick Riolo.

In 1986 the BACH group spent two days in Cambridge, Mass. as guests of Polaroid inventor Edwin Land at his Rowland Institute and of Danny Hillis, who at Thinking Machines Incorporated, was developing The Connection Machine, a massively parallel computer just perfect for genetic algorithms and agent-based models. Right after this trip, Carl Simon coordinated a BACH proposal to the NSF for funds to bring a Connection Machine to UM. UM President James Duderstadt promised matching funds from his own budget, and the NSF approved the application, but within the next month withdrew the approval because it had to deal with unrelated financing irregularities in the awarding division.

In 1987, the BACH group celebrated Bob Axelrod's MacArthur "genius award"; in 1992, it celebrated John Holland's.

In the mid-1990s, the BACH group faced a crisis when the Santa Fe Institute tried to recruit John Holland to leave UM. Partly through the efforts of John Holland, as documented in Mitch Waldrop's book *Complexity*, the Santa Fe Institute (SFI) was founded in the mid-1980s as an interdisciplinary think-tank on complex systems. Axelrod, Cohen, and Riolo also played roles in those formative SFI years.

When SFI invited John Holland to join their core group of resident faculty, the BACH group schemed long and hard on how to keep Holland in Michigan. With the advice and encouragement of the BACH group, Holland negotiated an agreement with UM LS&A Dean Edie Goldenberg and Business School Dean Joe White in which Holland would stay at UM with LS&A and the Business School paying for a workshop every year in which SFI faculty would spend a week at UM sharing ideas with select UM faculty as invited by the BACH group.

II. PSCS (1995-1999)

Spurred on by the success of these UM-SFI meetings and the growing recognition of the importance of complexity in science research, the UM Vice President for Research (VPR) Homer Neal began regularizing support for complex systems at Michigan and in 1995 officially established the UM Program for the Study of Complex Systems (PSCS), with physicist Robert Savit as Director. Its main activities were the UM-SFI workshop every Fall and the UM Complexity Conference every March to acquaint UM faculty with the ideas and techniques of complexity.

PSCS Director Savit was able to procure office space for PSCS, funds for an administrator, and a computer lab and funding for Rick Riolo, who was appointed Assistant Research Scientist of Complex Systems. During these PSCS years, Simon began the Nobel Symposium in which UM faculty would summarize the lives and research of each year's Nobel laureates. The Complex Systems Rackham Certificate Program was also started; graduate students who took a five-course program in complex systems, including Rick Riolo's Introduction to Complexity Course and Agent-based Modeling Course, would receive recognition of their efforts and expertise on their UM diploma. The Complex Systems Certificate remains one of the most successful graduate certificate programs at the university, averaging about nine students per year.

Within two years of the founding of PSCS, the university set up an external review committee to examine the past and future of complex systems at The University of Michigan. SFI President Ellen Goldberg chaired the committee. The committee's report was enthusiastic and urged the university to deepen and regularize its commitment to complex systems and to PSCS. UM Provost Nancy Cantor and Assistant Provost Paul Courant were convinced and proposed regular funding and some faculty lines.

In May of 1999, during that transition period, PSCS held a major week-long conference to celebrate John Holland's seventieth birthday. The Festschrift was organized by four of John's Ph.D. students: Lashon Booker, Stephanie Forrest, Melanie Mitchell and Rick Riolo. Speakers included Economics Nobelist Ken Arrow and Physics Nobelist Murray Gell-Mann.

III. CSCS

And so, in 1999, with this new funding and hiring structure in place, The Program for the Study of Complex Systems became the Center for the Study of Complex Systems (CSCS), with Carl Simon as Director. It had a $600K annual budget, with $150K coming from the Provost, $150K from OVPR and $75K each coming from the Deans of Medicine, Engineering, LS&A, and Business. In addition, CSCS had approval to hire six new faculty — all with split appointments with other UM units.

CSCS made the following faculty appointments in its first ten years:

- 2000: Scott Page (1/2 CSCS; ½ Political Science)
- 2001: Mercedes Pascual (1.0 Biology; 0.0 CSCS). After the joint search and hiring process, Biology Chair Julian Adams decided he did not want to split Mercedes' appointment. She is still very active in CSCS.
- 2002: Mark Newman (1/2 CSCS; ½ Physics)
- 2005: Lada Adamic (1/3 CSCS; 2/3 Information) Lada is now on leave to Facebook
- 2007: Robert Deegan (1/2 CSCS; ½ Physics)
- 2008: Elizabeth Bruch (1/4 CSCS; ¾ Sociology)
- 2008: Charles Doering (1/2 CSCS; ½ Math). Doering moved his half of his full professor position to CSCS as part of his retention process.
- 2009: Pej Rohani (1/2 CSCS; ½ EEB). Hired as full professor.

The first six were hired as assistant professors and are now tenured; Bruch's tenure decision will be in 2014. Doering and Rohani moved to CSCS as full professors. Page, Pascual, and Newman were awarded Collegiate Chairs in 2008, renewed in 2013. Mercedes Pascual was awarded a Howard Hughes Fellowship in 2008. Bob Axelrod was appointed the Walgreen Professor in 2006.

116

In 2003, CSCS scored a major funding success when it was awarded a $3.4M IGERT grant from the National Science Foundation in 2003. The grant provided two years of support for Ph.D. students who were writing dissertations that use complex systems techniques to study economic and political institutions and whose thesis committee includes a CSCS faculty member. Forty-one Ph.D. students were supported by this grant between 2003 and 2011.

In 2005, CSCS morphed from an incubator unit under the Vice President for Research to a fully funded unit within The College of LS&A. In 2009 the Dean of LS&A appointed Scott Page as the new CSCS director. In these last four years, CSCS has further developed its undergraduate offerings, including four new courses and a complex systems minor. The minor has grown from just two students in its inaugural year of 2010 to its current enrollment of 34 students. Most recently Scott Page taught one of the first MOOCs (Massive Open Online Course) on Complex Systems Modeling, attracting over 40,000 online students. The course continues to be offered as one of the first online courses developed by the University of Michigan.

CSCS has now earned an "enhanced program status," which allows it to grant tenure. It holds a unique position among complexity centers in the world in that it includes all disciplines, as the above hiring list indicates, and has active undergraduate and graduate education programs. Throughout all these processes there has been one important constant: the guiding hand, inspiring imagination, and keen intellect of John Holland.

IV. AFTERWORD: Where are they now?

Doug Hofstadter is the College of Arts and Sciences Distinguished Professor of Cognitive Science and Computer Science at Indiana University.

Melanie Mitchell is Professor of Computer Science at Portland State University.

Michael Savageau is Distinguished Professor of Biomedical Engineering at UC Davis.

Sadly, Arthur Burks died in Ann Arbor in 2008, and Michael Cohen died in Ann Arbor in 2013.

William Hamilton moved to Oxford in 1994 to be The Royal Society Research Professor in the Department of Zoology. He died in 2000 from an illness contracted in a field trip to the Congo to investigate the origins of HIV. He had expressed interest in returning to The University of Michigan and CSCS. He published his reminisces of his first meetings with Axelrod and his first BACH interactions at the beginning of Chapter 4 of Volume 2 of *Narrow Roads of Gene Land* (Oxford, 2009).

Robert Axelrod, Rick Riolo, Carl Simon, Scott Page, Mark Newman, Mercedes Pascual, and, of course, John Holland are still actively researching and teaching on complex systems at The University of Michigan.

The BACH group in the late 1980s. From left to right: Michael Cohen, Robert Axelrod, William Hamilton, Arthur Burks, John Holland, Rick Riolo, Michael Savageau, Carl Simon (photo from Carl Simon).

The BACH group in the mid-2000s: John Holland, Scott Page, Michael Cohen, Robert Axelrod (holding the late 1980s group photo), Rick Riolo, Mark Newman, Carl Simon, Mercedes Pascual (photo from Carl Simon).

30   Qing Tian

Department of Computational Social Science
Krasnow Institute for Advanced Study
George Mason University

The Inspiration of John Holland

When I moved to Ann Arbor to pursue my Ph.D. study at the University of Michigan (UM) in 2006, I did not know who John Holland was and had no idea about complex adaptive systems (CAS). In my first summer at Ann Arbor, my adviser Dr. Dan Brown recommended to me Mitchell Waldrop's book, *Complexity – Emerging Science at the Edge of Order and Chaos*. I read it and instantly fell in love with CAS. I have always thought any science, to be a good science, must have something to say about the perpetual question of being and meaning. The science of CAS did just that. I still remember vividly from the book that as a group of brilliant people gathered in SFI and were grappling toward a new science, the Master of the Game stood up and explained the crucial properties of CAS with such clarity and elegance. Among the audience the Irish Hero Brian Arthur listened hard and scribbled furiously in his notebook. To me that was the highest moment in the book.

My first meeting with John Holland happened in my second semester at UM. It was on New Year's Day. While I was walking along the Huron River, it came to me that I could talk to him about my ideas of sustainability and CAS now that we were at the same university. Perhaps only after I had built my first agent-based model and tested myself in classes with subject matters that were completely new to me, did I dare to think so. And, of course, I was curious about the Master of the Game. UM has a program called "Take a Professor to Lunch." Each semester the university pays a number of students to have lunch with their favorite professors. The purpose is to foster interactions between students and professors. I decided to take advantage of the program. So I emailed John Holland about my new year's wish to take him to lunch. His reply came back quickly, suggesting we meet before or after his class *CAS and Emergence*. I went to his first class and sat there throughout the whole semester.

In the next five years during my study at UM, John Holland's pioneering work in CAS had not only inspired my thinking on coupled human-environment systems (CHES) and sustainability — the topic of my dissertation — he was also a constant source of encouragement throughout my dissertation work. I looked at CHES as CAS and explained sustainability as a global property emerging from actions and interactions of multiple human agents under the large social, economic, and institutional setting and interactions between the human system and the natural system. I proposed a framework that combines agent-based modeling with other research methods (GIS, remote sensing, social surveys, and interviews) to measure and assess the well-being of a CHES, to understand how complex interactions in the system shape its well-being and to explore possible future paths of the system. I applied the framework to a case study in the Poyang Lake Region of China on rural development amid flood hazards. I built an agent-based model to explore the potential effects of an alternative policy that subsidizes rural households that rent out land use rights for long terms in comparison with the current policy of subsidizing rice cultivation. John Holland took great interest in this work. He read and commented on my manuscripts. He even followed every major step of the development of the agent-based model and contributed to some of the ideas implemented in the model.

Near the end of my study at UM, I attended a meeting at SFI on stability of economic systems, fortunately with John Holland. It was my first visit to SFI. I got to meet some other legends in Waldrop's complexity world, such as Murray Gell-Mann. Although things may have changed since SFI first started, much of the spirit remains. The interest in CAS shared by all the participants was evidently strong. I read Waldrop's book again and realized that Murray Gell-Mann had a vision on sustainability even back then. In his vision a sustainable human society is adaptable, robust, and resilient to lesser disasters, can learn from mistakes and allows for growth in the quality of human life instead of just the quantity of it. He said the transformation to a sustainable society requires understanding economic, social, and political forces that are deeply intertwined. That was exactly what I was trying to do in my dissertation work. From Waldrop's book I also rediscovered Brian Arthur's insightful remarks on policy which would become part of a new course I would teach on *CAS and Policy* at George Mason University later. A loop thus closed and expanded.

Through our many interactions, I got to know some other sides of John Holland. John Holland is, first of all, an optimist. His optimism is constantly manifested in his bright smile and cheerful joking. In his quick steps rushing to pick up the first Mayapple on the ground and climb up boulders to watch the Potomac, I see a deep curiosity in him, the same kind of curiosity that has driven his scientific research. In his urgent tone of speaking, I hear an enthusiasm, with which he pursues science and approaches life in general. And he likes to play. He finds fun in plenty of things in life: flying a dragonfly in the field, picking up wild raspberries from bushes, spotting a Lady's Slipper or Indian Paintbrush on the roadside ... Only such curiosity, enthusiasm, and playfulness from within can sustain the kind of efforts that are needed to teach and write in one's eighties. And this curiosity, enthusiasm, and playfulness, I think, distinguish true scientists from career climbers and great scientists from good ones. John Holland is truly a scientist and a great one.

Today I am no longer in Ann Arbor, Michigan, where my mind underwent a transformation through interactions also with many other great people in the complexity program. John Holland and the science of CAS he and others have started, however, will continue to be a source of inspiration as I pursue my academic career.

Note: Included are two small pieces I wrote during my study at UM and a few email exchanges with John Holland on agent-based modeling. They sort of provide bits of evidence of John Holland's inspiration.

Appendix I. Summer 2006 – First Encounter

It is my day again. But I am no longer in the suburb of Washington DC, where I used to weave C++ code into a piece of software and would take off from work on this day. I would buy flowers, lots of them, from the farmer's market in the early morning and arrange them carefully on my balcony. I would sit on the balcony looking into the lush woods and feel spring springing in the air. I would watch squirrels chase around cheerfully under the sun. I would let my thoughts drift, wondering about the perpetual questions of "being" — a human is born to this world without his/her own choice, is there a purpose? What is the purpose? And what is it to live?

Now I am in school. I am reading Mitchell Waldrop's book, *Complexity – Emerging Science at the Edge of Order and Chaos.* What beautiful minds they have, John Holland, Chris Langton, Stuart Kauffman, Brian Arthur! Complex Adaptive System (CAS), such a beautiful idea! "A dynamic network of agents," "interacting with and adapting to each other and the changing environment locally," "coherent behaviors or global properties of the system arise only from micro-level actions and interactions in a bottom-up fashion," "no central control," "no equilibrium," "non-linearity," "perpetual novelty," "the whole is larger than the sum of its parts." Stock markets, social insect colonies, the biosphere and ecosystems, body cells, the brain, the immune system, any human cultural and social societies all are complex adaptive systems! Suddenly Physics, Biology, Ecology, and Economics all come together under a much larger science. It seems as if the whole world had changed, but what has changed was just a view of looking at the world. "...A point of view? One with the twister in vista glide ..." John Holland quoted the poet Alice Fulton in his book Hidden Order, illustrating CAS to a general audience.

I am amazed, thrilled, overwhelmed. On my walk along the Huron River in the afternoon, my spirits soar high above the clouds all the way. Thoughts hit upon me from here and there. All those questions that I have been asking myself for so many years, to which I have tried to look for answers from Socrates, Plato, Schopenhauer, Kant, Locke, Hegel, Nietzsche, Kierkegaard, Lao Zi, and Zhuang Zi, I realize, can be explained by science too, by this new science of complexity.

If the edge of chaos is where the complexity lies, where the system can do things, hence where the system is alive, it can well be the system is all about. The more complex it is, perhaps there is more meaning to it. Could this be true for being? Is being nothing but experiences? The richer the experience, the more meaningful a life is?

Fate is nothing but a path a human being makes for one's life. There could be infinite number of paths for each life, but at the end only one is walked. It is not because this is the optimal one, but it is what the individual can make based on his/her "internal models" interacting with others and the changing environments where accidents are ubiquitous. This path may be comprised by many short lines with one connected to another. Where the path has begun and how the earlier ones have gone might well shape how the later ones go ("path-dependence"). While the path of an easy life may appear as a simple straight line, the struggling of a lofty mind can write a very irregular curve. The complexity of the path, not where the path ends, tells the richness of experiences.

We human beings are proud of ourselves being thinking Homo sapiens, rationally following social customs, political orders, and economic rules. We have evolved into such a highly ordered society that all one does is to find a position in the big production machine, stay there and keep running. The mass are performing everyday this way. Stability, as John Holland said, is death, so the mass are dead. Isn't it ironical that we are becoming dead because of our intelligence?

Even though human societies have evolved into a highly ordered state, it doesn't ensure we are going to have this security forever. As one of many complex adaptive systems in a much larger complex system, the universe, we have to co-live and co-evolve with many others. Will global environmental change alter human destiny? James Lovelock has claimed that the world has passed the point of no return for climate change, and civilization is unlikely to survive. Will the energy crisis transform the human landscape into chaos? Gasoline price has been rising; oil production has reached the peak; it seems the time is coming that there won't be enough energy supply for those big suburban houses and SUVs. If this happens, will another "self-organization" start again? Will it be a better world after self-reorganization? Who will be the new world power? Will there be humans at all? Nothing seems impossible in a vast space of possibilities.

Nature is beautiful because it is not completely settled into order, neither is it in total chaos. At the edge of chaos, new lives keep emerging while the old die out. That's where the loveliness comes from. It is constantly changing and full of surprises.

Nature is beautiful, but we are destroying our beautiful home at an ever increasing speed. Urbanization has taken up more and more forests and precious croplands; wild life species go extinct faster due to habitat loss associated with human activities; fisheries have collapsed in many oceans because of over fishing; water resources have been contaminated by pollutions; severe natural hazards occur more frequently as a result of man-made climate change. What is sustainability then?

Sustainability or unsustainability is but a global property of coupled human-environment systems emergent from the actions of humans and laws of nature, the interactions between humans and nature, and the long term feedbacks. Does sustainability mean that humanity and nature get to a state of equilibrium? Is it possible to achieve such a state? The new science of complexity tells us it's meaningless to talk about equilibrium in a complex adaptive system as the system is constantly in flux and in "perpetual novelty". Then what can we do? Can we really do anything? Maybe the best we can try is to better understand ourselves and our partners and the interactions, respect each other as equals, so the game can at least continue even if we can't play together happily.

...

When I finish up the last page lifting my eyes from Waldrop's book, right outside the window, the sun is setting with a wholly new color palette and cloud pattern. I run downstairs to replace my old car license plate with the Michigan one — I am so glad that I have chosen to come back to school. It's the first time in my life I have become excited about a science. It's the first time on this day things seem to become a little clear. I even forget to eat my long-life noodles, the ritual I usually perform on my day. Guess it doesn't matter, if being is not about how long a life lasts, but "hanging on at the edge of chaos, be alive." I feel alive.

Appendix II. Fall 2010 – A loop closes and expands

So, I have made my pilgrimage to SFI
A bright place on a bright hill, from where
The bright ideas of some bright people
Have brightened my mind
As a Buddhist to Bodh Gaya
A Muslim to Mecca
A Christian to Jerusalem

Sitting between the fathers of CAS, in a conference room
I become lost in some kind of aura
While the father of Quarks quickly relates my Chinese name to
Japanese
"It's a square plus a cross," he says
The father of Genetic Algorithm jokes (as he always does)
"She is a peasant." (Tian means agricultural field)
But the father of Q says I am a good-looking peasant (which makes
me laugh)

It seemed as if just yesterday
Those legends in Waldrop's Complexity world gathered here
Launching a scientific journey on CAS
But when I watch Doyne Farmer (one of the legends) on the podium
Who is talking about regulation and stability of the economy
I have to believe
Twenty five years have passed

In twenty five years, people do age, and things change
But the spirit remains
The father of GA still has the highest volume of voice in the room
His remarks are sharp (just as his eyes)
The father of Q still has beautiful dense curly hair (despite it is gray)
His head is up and face alive when it comes across names
Like Faraday, Maxwell, Albert

Not just them though
The Yale economist, the Stanford demographer, the UCLA
neurologist
The Harvard biologist, the UM archeologist...
The staff members, even the waitress in a restaurant
And those businessmen from Intel, Lockheed, Citicorp...
All are enthusiastic (and interesting too) –
Enthusiastic about CAS

So twenty five years have passed, how much progress have we
made? I ask
It's very little, the father of Q says
We need LOTS LOTS of people to work on it, he continues
And we need theorists besides data mining, the father of GA adds
Let us get on and carry on what the legends have begun
Let us march on, as we must
No matter how hard it is

P.S.

Retracing footprints of the legends
I drive up the mesa from Santa Fe to Los Alamos
Stopping at the Valley of the Rio Grande, I watch
Sangre de Cristo Mountains above and far
As the legends did
The father of GA tells me
Sangre means blood and Rio is river

Rereading Waldrop's Complexity
I realize
Twenty five years ago, the father of Q
Who helped set up World Resources Institute
already had a vision on sustainability
which is exactly what my dissertation is about
I must study hard to get my "union card"

Luckily sitting next to Cormac McCarthy at breakfast
I get to have a small conversation with the Pulitzer winner writer
who said it is more important to be good
than it is to be smart
which I can't agree more
And he looks genuine and surprisingly gentle
Thus my pilgrimage to SFI goes beyond satisfaction

Appendix III. On Agent-based Modeling: Email Exchanges with John Holland

03/17/2007
Dear professor,
Thank you very much for the nice comments on my paper. I was very excited when I was writing it. But when I thought about it again, I started to have a mixed feeling. This mixed feeling is not just about the model I built but also about the modeling approach in general. How could I validate my model? Did not I interpret the model results in an overly stretched way? Somehow I feel I got those nice insights from the model only because I had them in my mind, and I then chose the factors (only part of the system) and a representation (one of many possible representations of the same system) and constructed the model in a way that could prove those subjective points of mine.

In general, how do we know that a model really captures the key mechanism in this system (comparing the generated patterns with observed patterns in the real world does not tell anything as we know that we could produce the same pattern by making up totally different mechanisms)? In addition there are so many possible representations of the same system in models, which could all generate the same patterns and people usually only model part of the system that favors their points.

For instance, I am sure somebody who has totally different views than those of mine, could build a model that also gives rise to cities/towns and proves his points as well. Actually the readings from Rick's class were all about this matter and came right after I started to think about these issues. Some people have expressed their deep doubts on modeling in their papers. These are the questions they asked:
Are we pretending to do what cannot be done?
Are we trying to predict the unpredictable?
After all, what good are models?
Qing

03/18/2007
Dear Qing,
Model validation is, of course, an important question: Are the mechanisms postulated the ones that are really operating? Even Maxwell's equations require validation, and the mechanisms

postulated change over time. We go from fields, to photons exchanging energy, to quantum interactions, and so on. The interpretation determines what we consider as validation.

Here we're back to the purpose of the model. If it is "data-driven" then prediction is the key. That has been the strength of Newton, Maxwell, and Einstein in the equation-based models they propose. However, in some cases, it is extremely difficult to think of ANY set of mechanisms that will produce the desired outcome. That was the strength of von Neumann's mechanistic model of self-reproduction — no one had been able to exhibit a mechanistic model before that.

This is what we called an existence proof model in class. I think your model falls in this category — choosing relevant factors is part of the "art form" (and it depends upon your subjective insights). As with von N., you had an objective and you showed it could be implemented with certain mechanisms.

If there is a second set of mechanisms that will yield the same result, then the objective is to find "real world" experiments that will distinguish between the two models. That's the difference between the work of Einstein (theorist) and the work of Eddington (experimenter). Good models, like good theories, tell us where to make new observations in the real world. Even cas have repeating, controllable patterns. The object of theory is to help us to find them.

Now, what do you think about THIS tirade?

John

03/18/2007
Dear professor,
THIS tirade is well received. But I am not quite sure I understand "The interpretation determines what we consider as validation." Do you think every model should be validated in one way or another? Do you think generating patterns that match the real world is enough for validating my model?

Probably science, not just modeling, itself cannot be totally objective. Just came across an article for Rachel's class. Here are some quotes from it:

"Science is more art than truth, created by people, who operate, like all of us, in a conscious and unconscious universe. Guided by "inner voices", researchers are inspired to discovery."

"Scientists are more like artists, assembling and mixing the colors of an awesome, complex and dynamic story into a coherent picture. Data maybe the paint on their pallet that creates a picture, but, inevitably, what we see is the artists' rendition. Decisions are made at multiple steps along the way: on relevant facts; on the boundaries between what we know and don't know; and on what we care about."

"Data are not just data. Information is always accompanied by interpretation."

Remember what James said about perceptions: perceptions are never pure sensations but more of interpretations based on the past experiences.

So I guess it is OK to have subjectivity in models now that it is inevitable. Still I think we need to be careful about models: sometimes people tend to take it for granted without any validation. Even if it is an exploratory model, there should be a certain level of rigor in them.

Qing

03/19/2007
Dear Qing,

This is a good dialogue!

I try to think of the model as a kind of axiom system: First, I try to make the basis of the model (the axioms) as clear as possible. I actually try to write an explicit list of assumptions. Then I try to make sure that the construction adheres to just these assumptions and no others. This is hard, but possible.

The whole purpose of setting up axioms is to move all questions of interpretation to them. From that point onward, the rules of deduction, or the program, are a "mechanical" working out of consequences, with no interpretation involved in that part (unlike arguments of rhetoric and persuasion). That is what, in my mind, separates the scientific method from other methods (say, philosophical argument).

In short, when the "axiomatic" approach can be followed, the art and interpretative cleverness are concentrated in selecting the axioms. Then consequences are "proved" without resort to interpretation.

Note, however, that intuition usually guides us in what consequences we would LIKE to show. But you cannot "cheat" the deductive method — the consequences may, or may not, follow from the axioms chosen.

Do you agree?

John

P.S. The quotes you give, in my opinion, mix up these two aspects of science. There is certainly all sorts of interpretation and intuition in setting up the axioms (say, Maxwell's equation), but the deductive consequences are then fixed (and no amount of social opinion can change them).

03/20/2007
Reflection on what is a good agent-based model

I can't stop thinking about this business of model validation. Let us still use Professor John Holland's "axiom system." In my mind I see three types of axiom systems.

Not much is known about the processes of a system, but we observe some patterns at the macro level. The purpose is to explain mechanisms underlying those macro patterns. In this case, the modeler can list all the axioms, including his assumptions about the mechanism. It is all right even if he "manipulates" the axioms. As long as his model generates the observed patterns, he can say that the mechanism postulated is plausible, which can then guide where empirical studies look. The ant model and John Holland's language model (how grammars emerge and how languages evolve) fall into this category. I would think these models are so called "existence proof model."

The model is used to test/explore abstract ideas. The modeler believes that a system works in a certain way. He kind of "proves" his belief (intuition) by generating macro patterns observed in the real world using his model. The purpose of the model is, however, not to prove his belief on the mechanism but to illustrate further insight about the system. In this case, it is OK to move all the assumptions to axioms and then let the program work out, whatever results are. But there is a limitation: whatever additional insights drawn from the model are claims of the modeler based on his belief of how the system works. My model on Towns, Cities, and the Happiness of Humanity and some of the early social ABMs, like Axelrod's culture dissemination model, fall into this category. I tend to think such models are more about brain exercise to illustrate an insight.

The model simulates a real world system and has clear policy implications (such as Dan Brown *et al.*'s land-use models). In this case, only producing macro patterns that match real world observations is not sufficient. It is necessary to combine other empirical research methods to understand the essential elements and dynamics of the system, including how agents actually make decisions. In short, such models need to be validated at conceptual, micro, and macro levels. Of course, no model is a complete representation of the real world system, and it is impossible to fully validate a model. But we do need to have a high level of confidence on our models so we can convince policy makers — if we want to make a difference in the real world.

These three types of models are all useful. Because they are intended for different purposes, the requirements for validation are different too. While the first type does not need validation, the validation of the second type only involves matching macro patterns, and the third type requires validation at conceptual, micro, and macro levels. In addition, if a model can generate MULTIPLE macro patterns observed in the real world, its credibility is enhanced.

To make agent-based modeling a rigorous research tool, we should explicitly state all the assumptions we make in any agent-based models (just as mathematicians list the axioms) and discuss how they may affect the conclusions drawn from model experiments. For example, a common implicit assumption for many social ABMs is that agents can go through a very large number of steps in one model run. This could have important implications on model outcomes. One may think about Schelling's famous segregation model.

A Serious and Playful Professor

31 Jan W. Vasbinder

I met John at a conference about artificial intelligence in Stanford in 1984. John talked about genetic algorithms. I did not understand much of his talk and afterwards I looked him up and introduced myself with: "Professor Holland, I am Jan and I am from Holland. I did not understand a word of your talk". That was the start of an immediate friendship that deepened as time progressed and that has given me great happiness and inspiration.

At the time John and I met, I was the science counselor at the Dutch Embassy in Washington. That wonderful job enabled me to travel to any place in the USA where something interesting was happening. So obviously, following our 1984 meeting, I traveled to Ann Arbor, where I got my first exposure to complex adaptive systems.

We moved back to the Netherlands in the end of 1985. In the fall of 1988 John, Maurita and Manja visited Amsterdam. My wife Irit, two of my daughters, Dorit and Nienke, and I met them there. We had a great time, including dinner and a lot of giggling between the girls. We planned for Dorit and Nienke to visit Manja, and we knew this was only the beginning. John told me about SFI, were the excitement had started.

Dorit and Nienke stayed in Ann Arbor in the summer 1989. They remember that John took them to a baseball game and that he gave them a lesson in drinking: "I drink whisky and water... I get drunk; I drink brandy and water... I get drunk: I should cut down on the water."

One of the wonderful things about John is that you do not have to see him all the time to deepen your friendship or continue your conversation. In fact each time we meet we just seem to continue where we left off the time before, even if there are years in between. And that sometimes happened.

In 1998 I visited John in Ann Arbor with Theo Groen, my business partner. Theo and I had a company that focused on finding and developing new combinations of science, and we were in search for material for a book. John organized a dinner in his home with some of his friends and colleagues. Before dinner John talked about SFI.

Then and there the idea emerged to start a similar institute in Europe. After dinner all of us went upstairs to John's study. I forgot the topics we discussed, but I will never forget the discussion. Here were some of the sharpest and brightest men on the planet, with an exceptional capability to listen, discussing some of the most interesting problems of our time. Later John told me that this type of discussions took place in SFI all the time. The bug that was going to change my life was planted.

Some years later the chief R&D officer of AKZO Nobel called to tell that the keynote speaker for an in-house conference on innovation had canceled, six weeks before the conference. He asked if we could arrange for a worthy replacement. I asked John. He came and stayed with us for the duration of the conference and a few days beyond that. We had a great time. One evening we had dinner with a few leading Dutch scientists. I asked John to talk about SFI, hoping he could convey his enthusiasm to a typical group of cynical Dutch professors, who asked questions like: "If SFI is so fantastic, why is there only one such place in the US?"

John in Holland, inside the IPL building, 2007

John did convince them that SFI was a unique place, and he answered that question, essentially as follows:

About a thousand plus senior scientists in the US have Nobel Prize like qualities. That is the quality you need to make a place like SFI work. That, and the capability to listen and be genuinely interested to find out what top scientists from other disciplines have to offer. Only 10% of the best scientists have that capability. That limits the population from which SFI can draw to about one hundred plus senior scientists. In Europe that situation is similar, so there is place for one SFI like institute in Europe.

John very strongly supported the idea to develop such a

European Institute, and when the initiative finally took shape John played a key role in getting the right people involved. He called Manfred Eigen, got his support and introduced me to him. Manfred subsequently made a list of 15 top-scientists, including five Nobel Laureates, called them all and introduced me to them.

When Institute Para Limes (IPL) was founded in August 2005, all the scientists suggested by Manfred plus ten others, including John, became founding fathers. John was there during the founding conference. So was Brian Arthur, who had picked up the rumor at IIASA and wanted to be part of it.

John could not attend the meeting in Strasbourg in March 2006 to formally start IPL, but Brian, Manfred, and quite a few others did.

But John was present at the workshop in 2007 on *Conceptual Neuroscience* and was one of the keynote speakers at the first international IPL conference (*Science without Boundaries*) in October 2007 in the Netherlands.

At that time John was already working with Helena Hong Gao (assistant professor at NTU) on language and complexity. At his suggestion I met Helena in Singapore to ask her to speak at the conference in the Netherlands.

John at his 80th birthday party

135

That meeting had far reaching consequences. During the meeting Helena mentioned that John would be 80 years in 2009. That triggered the idea for a conference in tribute to John, to be organized by IPL and SFI and to be hosted by NTU. Bertil Andersson, one of the founding fathers of IPL and provost of NTU (Bertil later became president of NTU), agreed and the conference *Adaptation, Order and Emergence* took place in February 2009.

We invited many leaders in the field of interdisciplinary science to speak.

Peter Ho, at that time the head of civil service in Singapore, was a guest of honor. As Peter later remarked: "That kind of hooked me into the nascent complexity community." That and the fact that the conference infected the top of NTU with the interdisciplinarity / complexity bug led to the establishment of the complexity program in August 2011 and to my appointment as its director.

John became one of the members of the Advisory Board.

In March 2013, at our conference *A Crude Look at the Whole,* John gave one of the best talks I have ever heard him give. The talk was about the future of the science of complex adaptive systems, the science that he fathered. The title was *Signals and Boundaries*, also the title of his (at that time) latest book.

While all this happened, I visited John many times in Ann Arbor, and John visited Holland and Singapore many times. We shared very happy times and some very sad. We always looked forward to the next time.

Some years ago in a reflective moment, I thought about the people that had the biggest influence on my life. John is in the top.

John changed my life, John changed a lot of lives, John changed computer science, John established complexity, John changed government agencies and in doing it all, John has been a great friend to many people.

I feel myself very lucky to be one of them.

32   Mitchell Waldrop

The Youngest Man

It's been more than two decades since *Complexity* came out, and longer than that since I started interviewing John Holland for details about his work and his involvement with the Santa Fe Institute. In that time, the details of our many meetings and phone calls have faded. But not so the memory of his unfailing good cheer and generosity in putting up with my incessant questions — nor the memory of the three qualities that make him so remarkable.

First, John is one of the most insightful scientists I've ever met. I deliberately choose the word 'insightful' rather than 'clever' — although he is that — because the world is full of smart people who can figure out the answer to any question you give them. But it's much rarer to find someone who can recognize that a question is there to be asked. John has demonstrated that capacity again and again in his explorations of genetic algorithms, classifier systems and more.

Second, John is one of the youngest people I know. Not chronologically, of course, but in his sense of wonder, his boundless curiosity, and his delight at understanding something new. It's the quality that Richard Feynman called 'the joy of figuring things out' — and, I think, an important part of what makes him so insightful. To John, it's just *fun*. When I think of him, what I remember most vividly is the laughing way he exclaims 'aha!' whenever he understands something.

Finally, John is one of the most open people I know. He eagerly listens to what you have to say as if he's really learning something new. Again, I suspect that that is an important part of what makes him so insightful: he seems to be willing to take ideas from everyone and everywhere. And in return, he's happy to share his own. I've talked to many leading scientists over the years, and know that many of them will leave you feeling impressed, informed, and sometimes even dazzled. But talking with John always leaves you feeling smart.

33   William S-Y. Wang

John and Me

John and I both did our Ph.D.'s in the 1950s at the University of Michigan. Our teachers, Arthur Burks and Gordon Peterson, were good friends. But we did not get to know each other well until an interdisciplinary Program in Communication Sciences was set up, and we were invited by our teachers to jointly teach the first course in the new Program. For my half, I talked about natural languages to a group of bright-eyed students coming from very diverse backgrounds. I still have fond memories of sitting in on John's elegant lectures, largely on automata theory and other aspects of computation. The experience reinforced in me the conviction that the fundamental questions in science must be examined from the integrated viewpoints of many disciplines, drawing knowledge from each as needed.

After I left Ann Arbor, we lost touch with each other. One day decades later, while I was browsing in Cody's bookstore in Berkeley, I was attracted by a little book called "Hidden Order: how Adaptation Builds Complexity". It seems a particularly cogent way of viewing how language emerged — from simple vocalizations and iconic gestures step of millennia ago building step by step into the remarkably complex systems we have today. I was overjoyed to see the author was John.

I couldn't be sure that John would remember me, but I emailed him anyway. This rekindled an old friendship, recounted in an interview reported in the Spring 2005 issue of the Santa Fe Institute Bulletin, from which the following photo is taken.

Scientists Reunite After Four Decades

"So," the reporter wrote, "43 years after co-teaching the first course in a field that would one day become complexity science, Holland and Wang co-taught a course in language evolution at the SFI Summer School in Qingdao." By that time I have retired from Berkeley and returned to China, situating myself in Hong Kong. John and I started to visit back and forth, sometimes at Santa Fe, and sometimes in Hong Kong, shown respectively in the two photos below:

I particularly like the one taken in Santa Fe because of the halo effect the sun's ray produced with my thinning hair. Both pictures show John with his broad infectious smile. As John said in that interview, "If we demonstrate that language can be learned with more primitive abilities, then that would change the way linguistic research is done." Both parts of his sentence are coming true.

It is an evolutionary perspective I strongly share. I could never understand how proposals of a 'language organ' can be taken seriously by anyone considering language emergence. In fact, I once joked about an "Organum ex machina" in a commentary for _Behavioral and Brain Sciences_ (1984), since the proposals reminded me of the "Deus ex machine" used in Greek tragedies.

Millions of years of biological evolution endowed us with the ensemble of neuroanatomical systems of respiration, mastication, sensorimotor control, memory, and cognition, and "primitive abilities" from each of these systems somehow got interfaced for language to emerge. Language made possible cultural evolution in our species, which in turn completely changed the face of our planet. A fundamental agenda for research on our species is to understand what these "primitive abilities" are, and how they came to interface with each other.

In recent years, we started a series in China, called Conference in Evolutionary Linguistics (CIEL), and John came to Beijing to give a keynote at CIEL-4 in 2012. CIEL-6 will take place in Fall 2014 in the beautiful coastal city of Xiamen, and we hope John will come again to share his wisdom.

Even more recently, a Joint Research Center for Language and Human Complexity was established among Peking University, the University System of Taiwan, and the Chinese University of Hong Kong, with the latter serving as its hub. Our aim is to "change the way linguistic research is done", in John's words. In place of making up complicated analyses of individual sentences in familiar languages, we will base our research on two empirical foundations. On the one hand, we will continue the time-honored tradition of field linguistics and investigate in depth the diverse languages and cultures of Chinese minorities. On the other hand, we will connect the field data with laboratory experimentation, especially examining how different languages shape different perceptions and behaviors, using the most powerful technology on brain imaging current available.

Writing was invented only several thousand years ago, enabling information communicated in language to accumulate and to be used across time and space. With the coming of the electronic age, information is building up at an ever accelerating pace and shared across the world at lightning speed. The complexities of our world, whether at the level of individuals, regional communities, or continental civilizations, are growing explosively. The world is certainly a very different place from when John and I taught our course in 1961. But the perspective of our course, couched in a perspective that is at once evolutionary and multidisciplinary, is even more relevant today.

I am particularly happy that the Nanyang Technological University offered the two of us an opportunity to get together again, this time in the dynamic city of Singapore. Lastly, here is a photo to remember the occasion by. Both the friends and the friendship have aged of course — all for the better, I think.

34   Tom Westerdale

The Holland, an Evolutionary Paradox

The holland is a friendly creature, often seen perched in high places overlooking lakes or river valleys. If we approach its roost quietly, we can glimpse it busily at work producing strings of text. To view this activity, it's best to approach in the morning. In the afternoon or evening, the roost is usually empty, for the holland is not a solitary creature. It is often seen in seminar rooms and classrooms.

The presence of a holland has a profound effect on all nearby organisms. Organisms that are near a holland feel good about themselves and about their own ideas.

What is unclear is how this fitness boost is achieved when all that seems to be happening is a lot of arguing and shouting.

The holland is a noisy creature. If we speak to it and say something like
((LAMBDA(X Y)(CONS Y (CONS X NIL)))(QUOTE TWO)(QUOTE FORTY)),
it is liable to respond by saying
011100000110100101111010011110100110001.
The holland's language has baffled evolutionary biologists, who find the phylogenetic position of the holland contradictory. It is well known that simplification is the essence of the evolutionary creative process, the streamlined Prokaryotes being the most advanced organisms. (See figure 1 and figure 2.) Thus the characteristics of the Holland would seem to mark it as a typically primitive organism. Yet its language, in its essential simplicity, seems to be one of the most highly evolved.

Its paradoxical phylogeny perhaps explains the paradoxical nature of the holland's communications, which often turn evolution on its head and talk of "the evolution of complexity".

Only as part of a general theory can the paradoxical holland be understood. Observations are useful, but are rarely described in a language that permits deductive exploration. A well designed mathematical model, on the other hand, generalizes the particulars.

Only mathematics can take us the full distance to an understanding of the holland.

THE MOST ADVANCED ORGANISMS ARE PROKARYOTES, NOT EUKARYOTES

primitive
(EUKARYOTE)
MUCH REDUNDANT DNA

advanced
(PROKARYOTE)
LITTLE REDUNDANT DNA

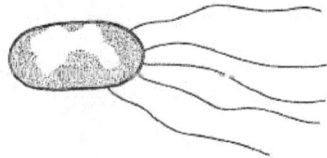

Ioannes bataviensis

Escherichia coli

Much Junk DNA
Introns
No Multi-gene Operons
Wasteful Diploidy
Separate Mitochondrial Genome
Carefree Smile

Whole Genome Useful
No Introns
Control through Multi-gene Operons
Efficient Haploidy
Single Streamlined Genome
Sober Countenance

Figure 1.

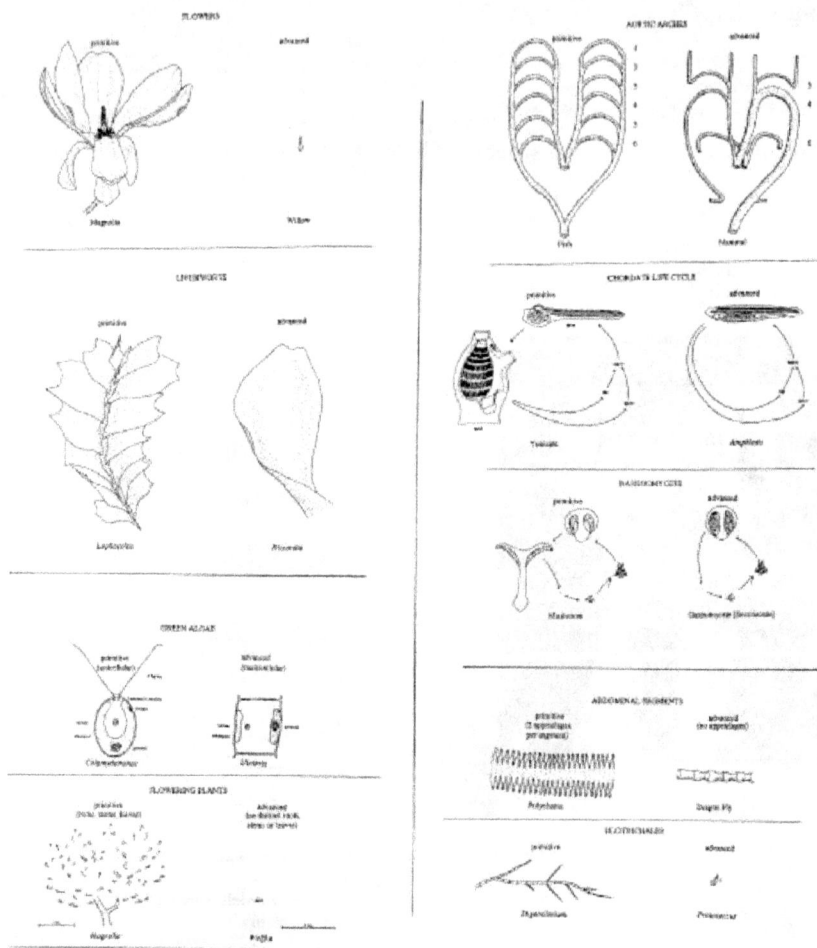

Figure 2.

35   Wei Zeng

John's STYLE

I feel extremely honored to contribute to this great book of John's stories. John is a very special person.

To me, John is more like a 'despite-of-age' friend than a prominent professor and great scientist. Considering all his fame and achievements, John is the most easy-going and understanding person I have ever met. No wonder he has so many friends of all ages all over the world.

It was the autumn of 2003 when I first met John, during a guest lecture he gave at Indiana University, Bloomington. At that time I was a graduate student in biophysics at IU. Since I have attended the 2001 Santa Fe Summer School, I knew John was a big name in the field of 'Complex System'. Often I felt very much stressed to speak to professors, but John's lecture was really interesting and the friendly way in which John delivered his lecture gave me some courage. I dared to come to the front after the lecture and had a short conversation with John. It was really fun and intriguing. I remembered John ended the conversation saying 'Email me and I will respond'.

I was able to spent considerable time with John and have a closer look at John during the next three years. At that time, I was at the end of 'PHD kindergarten' and the crossroad of life — writing the never-ending thesis without knowing what to do next. To quit science or not? Which country to stay? I was rather lost and upset. John comforted me by telling the story of his youngest daughter, who was also in the last year of her Ph.D. in Biosciences:

'She decided to take a break and step back to look at the big picture. It is very natural for young people to feel unsure from time to time. Give yourself some time.'

I did so. Before heading for UC San Francisco for a postdoctoral position in May 2005, I spent a few months exploring things I had always been curious about but never had the chance to give a try. It was at John's sunny house where I painted for the first time since high school. John handed me a watercolor set and said:

Recently I started taking watercolor classes and I found it really fun. Try.

I started with facsimileing a bird in John's painting textbook. The afternoon sunlight came down through the skylight. I was not good at facsimileing, so I soon turned into painting 'the bird' I had in mind instead. John loved the bird I painted and kept asking what kind of bird it was. John is very fond of animals and evolution. I really did not know. It is not a 'real' bird. I also painted some flower, fish, butterfly and bee. John framed all of them and hung them in his house. John did know how to make people feel appreciated. And during that 'dark period' it was very important for me to feel I was good at something. It is very kind of John.

(photo from Wei Zeng)

John is a good 'go' player and a very patient teacher. He could probably beat me with his eyes closed. But I was slightly better in Chinese chess, and John was surprisingly serious about losing the game. It was interesting to see the boyish side of a scientist. Maybe this is why he naturally gets along very well with his undergraduate and graduate students. I saw no generation gap at all. Actually John does have a GOOD reputation to 'spoil' his students and encourage them to pursue all kinds of 'crazy' ideas. His warmness, kindness, openness and wisdom make him a great advisor. During my postdoc days in San Francisco and 'post-science' days in China, when I feel lost, I always ask John for advice. And he was always there.

Once I showed John a picture I took at a castle in Finland, and he said:

'You captured the loneliness and stillness of time flow'.

(photo from Wei Zeng)

And this was exactly how I felt about this picture. It is great to find a mindset of similar kind. The intrinsic loneliness of scientists could be unbearable otherwise. Having a sensitive mind, John talked about foliage and snow, birding singing and frog in the pond. In addition to science, John's view of the world must be very artistic and poetic as well. On the other hand, John loves poker, chess, Lord of the Ring, Mason symbols etc. In poker game, he is known as 'the honest John'. And he ran away when the luck is not on his side. It is very rare to see a scientist of John's style, isn't it?

In my eyes, maybe also in many of his friends' eyes, John himself is a complex adaptive system that has evolved for more than 80 years. John could give you the feeling that he knows what you think but he would not speak it out. Frankly speaking, his brain still runs faster than those of many young people. John always spends a lot of time carefully preparing all his lectures and it is amazing how he could explain complex problems in a simple, elegant and clear way. In his 70 and 80s, John still does sailing and starts learning painting and cello, and still takes long business trips across the continents. John likes deep fried crispy duck, milk shake and dislike vegetables, but he is extremely healthy and energetic. This is against science. By the way, he does not cook.

John is also a philosopher and historian. He seems to have deep thoughts about most of the historical stories and happenings in the world. Among all the westerners I know, John has a surprising understanding and appreciation of the Eastern culture. His humor is cross-cultural. Even the shiest Chinese student found it joyful to talk to John. And John knows each of them. During John's several business trips to Beijing, I was astonished by how well he could fit into the environment. When we visited the museums, I was astonished by his knowledge of Chinese history and art. Of course he likes the Peking roasting duck, and he also loves the deep-fried Shanxi-style crispy duck. I heard some people saying 'John IS Chinese'.

John is very determined, and he fights hard for what he believes, both in research and in life. With parents doing business, John found his way from a talented and curious child to a great scientist. I am kind of 'jealous' of the joy John finds in his work. Bitter or sweet, up or down, John was lucky, in the sense of being always able to pursue his interest. To be the one you want to be, no matter how difficult, is what I have learned from John.

I hope I have line-drawn THE John I know of. As a rather maverick, I truly treasure the way John understands me and encourages me. Thank you, John.

36   Xiaomu Zhou

I've known about John since 1998 when I was a faculty member in the School of Information at Renmin University of China in Beijing. One day, my dean professor Yu Chen asked whether I would be interested in translating a book that he brought back from his recent tour to the US. It's written by John Holland, titled *Hidden Order: How Adaptation Builds Complexity.* Honestly, I had never thought whether I was even capable of such a thing, given that my English training had been all in China and I had never been to the US. However, Professor Chen encouraged me to read a few pages and then decide.

Then I was immediately drawn into the book, and was amazed how the concept of complex systems had been explained and interpreted via many academic disciplines, e.g. ecology, biology, economics, medicine, sociology, etc. This was a topic that most Chinese scholars had never been exposed to, and Professor Chen was especially good at recognizing the potential interest from Chinese scholars in this new interdisciplinary research topic. With his encouragement and trust, I started building blocks, one day a couple of pages, and finished my major translation within three months. We included a student (Jing Han) to help work on the last two chapters due to the scheduled publication timeline. As John later told me, based on his interaction with many Chinese scholars in complex system research, the Chinese version of the book had indeed played an important role in opening an interesting field for many Chinese, working in mathematics and other disciplines, who began to address issues within the complex system framework.

My real interaction with John did not start until later in 2000 when I received a visiting scholarship from the Chinese government that allowed me to travel overseas to learn the American education system. In fact, I simply did not remember that John was associated with the University of Michigan. During my time at the Bentley Historical Library, which was my host, John's name and his affiliation came across me one day. I just could not believe that was such a great coincidence. I immediately sent John an email, with an opening line, "I am one of the translators of your book Hidden Order......, and I am now in Ann Arbor."

Our friendship started from there. My own career path was later influenced by my experience in Ann Arbor, particularly with John and several other people, which led me to pursue a doctoral degree at the University of Michigan, and ultimately to a faculty position at Rutgers. During my years in Ann Arbor working on my doctoral degree, John kindly offered his house for my son, Yutong Chen (then 13 years old, and now a freshman in Carnegie Mellon University), and myself to visit for a year to save expenses. Through these years, we have also become good friends of John and his family members, and we often go back to Ann Arbor to visit John.

While everyone knows about John's pioneering work in complex system, Yutong and I have the luck to know more about him as a person in daily life and his outstanding knowledge of birds and mushrooms. During the year we stayed in his house, we tried different kinds of mushrooms grown around his house (including morels). John's way of treating these mushrooms is always the same, i.e. slightly sautéing the mushroom with butter and a bit salt, and then serving the mushrooms on a piece of fresh toast. This has become my family recipe when we have a chance to get fresh and delicious mushrooms. John's knowledge about birds is like an encyclopedia to us. Every time we run into a new bird, we often call John first to find out what this might be and then go online to search, and John is always correct.

John is lucky to have great genes. He can still be so active writing books in his 80s. He has been largely healthy compared to many other 80 year olds, even though he does not regularly exercise, he is very sedentary, and has just cold meat and bread for most lunches. While mostly working at home, John maintains a regular time schedule. When he writes, he is extremely focused. He once worked on his book for almost five hours without noticing the time passing, no bathroom trip, no standing up. Everyone at home thought he was not in the house, but he really just had to write non-stop in order to catch all his new ideas.

If he is not working on his books, John would spend time lying done on his couch to play poker with a small PDA, or read a science fiction novel. When friends or family members encourage him to exercise more, he has his own theory. John would defend his life style with turtle wisdom, that is, turtles are very sedentary so they preserve body organs from operating too much, and so they generally live the longest among all animals.

Perhaps the most interesting experience is to hear how John comments on the nerds and technology geeks portrayed in *Big Bang Theory*, and their behaviors. Clearly, John enjoys the show, but thinks those nerd geniuses are really out of touch. Ironically, as a big scientist who mostly lives in his own world thinking and writing, John can be easily seen as one of them, and many of his social behaviors are very interesting and unique.

Here is a story that John shared with us. In his first year at MIT for his undergrad degree, coming up to Valentine's Day, John walked over a couple of miles to find a butcher, asking for a real heart. The butcher gave John a pig's heart. John then mailed this bloody part to his high school sweet heart back in his hometown to give her his heart. How was this romantic action perceived? They broke up immediately. The interesting thing is that, after so many years, John seems still not to understand why the girl was scared, and acted in the way she did, and why his romantic message was not received.

While John can sometimes be misunderstood by others socially, he is in fact very personable, easy-going, agreeable, playful, charming, and humorous. Wherever John is involved, you can hear loud laughing from the crowd. Yutong and I are so blessed to have John as our friend and we always feel that John is like our family member.

The photo was taken in August, 2007, on the balcony of John's summer house in upper peninsula, Michigan (photo from Xiaomu Zhou).

37 Time Line of Big Events in John's Life so Far

2 Feb 1929	Born at Fort Wayne, Indiana. Raised in Western Ohio
	Parents: Gustave August Holland (Gust) and Mildred Prudence Gfroerer; One sister: Shirley (Holli)
10 months old	Spoke his first word at his grandmother's house: "shawl"
	High school (Van Wert Ohio). Because the school was so small, John had to tutor with the town's sanitation engineer to meet the trigonometry requirement for admission to MIT
1950	BSc, Physics, MIT
1950-1952	Member, Planning Group for IBM 701 Electronic Computer
1952-1956	Grad Student at MIT, Consultant on Problems of Concept-Forming Machines, IBM, to provide his support
1954	MA, Mathematics, University of Michigan (UMich)
1956-1958	Research Associate, UMich Institute of Science and Technology
1959	PhD, Computer Science, UMich. Officially the PhD was in 'Communication Sciences'; MIT and UM were the only universities that had such programs and in both cases they later became departments in computer science
1959-1961	Associate research Mathematician: Umich Institute of Science and Technology
1959-1964	Lecturer, Departments of Philosophy and Psychology, UMich
1959-1964	Chairman, Summer Conference Course, Programming Concepts, Automata, and Adaptive Systems
1961-1964	Assistant Professor, LSA College Appointment
1961-1964	Research Associate, Carnegie Institution of Washington, D.C.
1964-1967	Consultant, Argonne National Laboratory
1964-1967	Associate Professor of Computer and Communication Sciences, UMich
1965-1976	Executive Committee, Department of Computer and Communication Science

1965-1978	Policy Committee, Department of Computer and Communication Sciences
1967	Professor of Computer Science and Engineering, UMich
1975	Book: Adaptation in Natural and Artificial Systems
1976-1979	Council for International Exchange of Scholars, Advisory Screening Committee in Computer Science
1979-1982	University Senate Assembly, UMich
1979-1984	Policy Committee, Department of Computer and Communication Sciences
1980-1985	Executive Board, UMich Center for Cognitive Science
1983-1984	Executive Committee, Department of Computer and Communication Science
1983-1984	Deans' Ad Hoc Committee for Merger of Computer and Communication Sciences and Electrical and Computer Engineering
1983-1986	Senior Fellow, UMich Society of Fellows
1985	General Chair, International Conference on Genetic Algorithms and Their Applications, CMU
1985	Visiting Scientist, Rowland Institute for Science
1986	Book: (with Nisbett, Holyoak, Thagard), Induction: Processes of Inference, Learning, and Discovery
1986	Conference Chairman, Workshop on Classifier Systems and Parallelism, Rowland Institute for Science
1986	Visiting Scientist, University of Bergen (under sponsorship of the Norwegian Research Council)
1986-1987	Ulam Scholar' at the Center for Nonlinear Studies at Los Alamos. Meeting with George Cowan and Murray Gell-Mann and beginning of involvement with the founding of SFI.
1987	Co-chairman, Santa Fe Institute Workshop on Computational Biology
1987	General Chair, 2nd International Conference on Genetic Algorithms, MIT
1987-1993	Executive Committee, University of Michigan Press
1988	Professor of Psychology, University of Michigan
1988	Co-editor, special issue of Machine Learning (genetic algorithms)
1988	Santa Fe's Artificial Stock market
1988-1989	President's Life Sciences Commission, University of Michigan

1988-1990	Advisory Committee, Artificial Intelligence and Robotics, Canadian Institute for Advanced Research
1989	Organizing Committee, Third International Conference on Genetic Algorithms and Classifier Systems
1992	MacArthur Award
1995	Book: Hidden Order: How Adaptation Builds Complexity
1998	Book: Emergence: From Chaos to Order
2009	80th Birthday: Conference in his honor in Singapore
2012	Book: Signals and Boundaries
	Additional Current Appointments
	Board of Trustees, Santa Fe Institute
	External Professor, Santa Fe Institute
	Founding Board member, Institute Para Limes
	International Academic Advisory Committee' of AMSS, Chinese Academy, Beijing
	McDonnell Foundation, Standing Panel on Complexity Research (panel does no longer exist)
	The Land Institute, Advisory Team
	Plexus Institute, Advisory Board
	Guest Professor, Qingdao University (not active anymore)
	Executive Board, International Society for Genetic and Evolutionary Computation
	Executive Committee, Center for the Study of Complex Systems, UMich
	Director, University of Michigan/Santa Fe Institute Research Program (Program no longer active)
	Associate, Institute of Humanities, UMich
	Editorial Board, Complex Systems
	Editorial Board, Adaptive Behavior
	Editorial Board, Evolutionary Computation
	Editorial Board, Complexity
	Editorial Board, Computational Intelligence and Applications
	Advisory Board, Mind and Society
	Advisory Board, Quantitative Finance
	Advisory Board, Journal of Bio-Education
	Advisory Board, Journal of System Science and Complexity

John Holland was born in 1929 in Indiana and raised in western Ohio. At a young age John had a craving for knowledge. His mother taught him to play checkers at the age of 4, which had a lot to do with his lifelong interest in games and rule-based systems. Both his mother and father were avid readers.

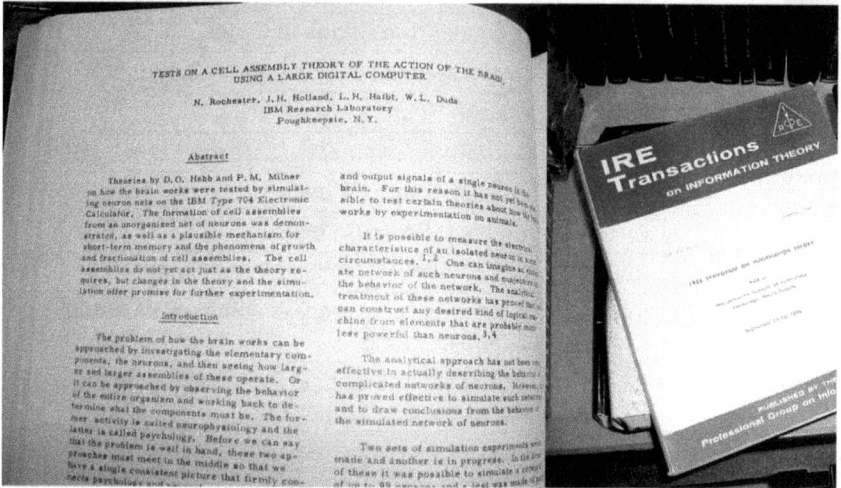

John's first paper in his life: Tests on a cell assembly theory of the action of the brain, using a large digital computer (1956) by N Rochester, J H Holland, L H Haibt, W L Duda.

Academically, physics and mathematics were his strengths. During his senior year in high school he took a state wide exam in these two subjects and missed first place by 2 points. He finished third and this got him a scholarship to MIT. This is where he began to explore the simulation of natural evolution to computers. "It would be twenty years before John Holland settled on it and twenty more years before people began to understand its significance." Holland received the first Ph.D. in Computer Science.

He was fascinated by a programming based on constricting artificial networks of metaphorical neurons (which is the idea that neurons came together to form a network from which created memories and complex behavior emerged). This worked well with Holland's ideas on artificial life.

John's B.S. thesis in Physics (required) was under Zdenek Kopal, an astronomer teaching 'numerical analysis' in Electrical Engineering, who gave me access to Whirlwind, the first 'real-time' computer (it even had a CRT display).

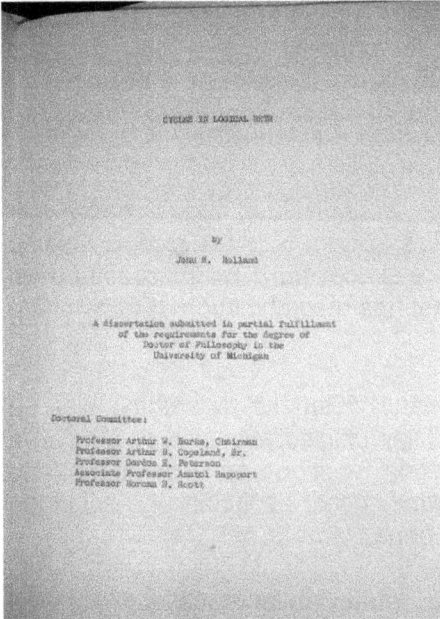

John's dissertation

Holland became an expert in computer programming and IBM asked him to work with an elite group of engineers, planning logical design of the company's first calculator the "701". To test the 701 they implemented a nerve-net system and used the computer as a lab rat. "Even then we understood there were real advantages of having these simulated test animals. The advantage was that we go inside and see individual neurons, start the thing over from the same initial conditions and go through a different training routine."

As Holland saw it, there was a link between biology and computation. Machines could be trained to adapt to surroundings the same way animals could. Bottom up: "Starts with a presenting a problem and virtual randomness and program nature into it." A book by R.A. Fisher, the famous British statistician, titled "The Genetical Theory of Natural Selection", changed Holland's life. The book saw evolution as an engine for adaptation. "Evolution was like learning a form of adapting to the environment.

John as a young faculty member at University of Michigan – photo from http://um2017.org/faculty-history/faculty/john-h-holland

It worked over generations, rather than a single life span." He thought if this theory could work so well for organisms, why not computer programs as well. This is where Holland imposed the theory of GA. "A Genetic Algorithm is a method of problem analysis based on Darwin's theory of natural selection.

It starts with an initial population of individual nodes, each with randomly generated characteristics. Each is evaluated by some method to see which ones are more successful. The successful ones are "mated" to produce a "child" that has a combination of traits of the parent's characteristics."

This was a brilliant step for Holland. "Genetic algorithms were a breakthrough in two respects: 1. They utilized evolution to provide a powerful way to perform optimization functions on a computer and 2. They provide window for the workings of evolution and a unique manner of studying natural phenomena."

From Genetic algorithms came what Holland called "schema theorem". Holland looked at "Fisher's theorem and saw it applied to individual genes. The schema theorem expanded how building blocks (combinations of genes) exerted their powers in GAS and indicated what might be a basis for population wide retention of combination of genes in natural biology." Schema is used to describe all strings that contain a given building block or set of building blocks. Proximity ('linkage' in genetics) plays an important role.

Holland was asked to be an external faculty member at the Santa Fe Institute (at that time there was no resident faculty). "The Santa Fe Institute is a private, non-profit, multidisciplinary research and education center, founded in1984. ...It has devoted itself to creating a new kind of scientific research community, pursuing the study of complexity and emergence".

38 Ph.D. Students under John's Supervision since 1965

(in some cases as co-chair)

Name	School	Year
Gul Agha	University of Michigan	1985
John Bagley	University of Michigan	1967
Albert Behtke	University of Michigan	
Theodore Belding	University of Michigan	xx
Tommaso Bersano-Begey	University of Michigan	2003
Lashon Booker	University of Michigan	1982
Ronald Brender	University of Michigan	1969
Daniel Cavicchio, Jr.	University of Michigan	1971
Edgar Codd	University of Michigan	1965
David Cohen	University of Michigan	
Clare Congdon	University of Michigan	1995
Kenneth De Jong	University of Michigan	1975
Marion Finley, Jr.	University of Michigan	1967
Stephanie Forrest	University of Michigan	1985
Daniel Frantz	University of Michigan	1972
Robert French	University of Michigan	1992
Leeann Fu	University of Michigan	2003
Andrew Gillies	University of Michigan	1985
Yehoshafat Give'on	University of Michigan	1966
David Goldberg	University of Michigan	1983
Michael Gordon	University of Michigan	1984
Paul Grosso	University of Michigan	1985
Stephen Hedetniemi	University of Michigan	1966

Roy Hollstein	University of Michigan	
Chien-Feng Huang	University of Michigan	2002
Dijia Huang	University of Michigan	1989
David Jefferson	University of Michigan	1969
Roberto Kampfner	University of Michigan	1981
John Koza	University of Michigan	1972
Michael Landy	University of Michigan	1981
Christopher Langton	University of Michigan	1991
James Levenick	University of Michigan	1985
Nancy Martin	University of Michigan	1973
Melanie Mitchell	University of Michigan	1990
Carl Page	University of Michigan	1965
Zollie Perry	University of Michigan	1984
Philip Pilgrim	University of Michigan	1975
Thomas Plum	University of Michigan	1972
William Rand	University of Michigan	2005
Robert Reynolds	University of Michigan	1979
Rick Riolo	University of Michigan	1988
Richard Rosenberg	University of Michigan	
Jeffrey Sampson	University of Michigan	1969
Michael Skolnick	University of Michigan	1984
Donald Stanat	University of Michigan	1966
Reiko Tanese	University of Michigan	1989
James Thatcher	University of Michigan	1965
Tommaso Toffoli	University of Michigan	1977
Roger Weinberg	University of Michigan	1970
Thomas Westerdale	University of Michigan	1969
David Wilkins	University of Michigan	1987
Annie Wu	University of Michigan	1995
Bernard Zeigler	University of Michigan	1968